Kazumi Tanuma

Stroh Formalism
and Rayleigh Waves

Previously published in the *Journal of Elasticity* Volume 89,
Issues 1–3, 2007

Springer

Kazumi Tanuma
Department of Mathematics
Graduate School of Engineering
Gunma University
Kiryu 376-8515, Japan

ISBN-978-90-481-7622-9
e-ISBN-978-1-4020-6389-3

Printed on acid-free paper.

9 8 7 6 5 4 3 2 1

springer.com

Contents

Foreword

Roger Fosdick

The *Journal of Elasticity: The Physical and Mathematical Science of Solids* invites*
expository articles from time-to-time in order to collect results in areas of research that have
made significant impact on the field of solid mechanics and that show continued interest in
present-day research and thinking. The Stroh formalism is one such area, especially as it
impacts upon the analysis of anisotropic media, that has been generalized from linear
elastostatics to cover linear theories of piezoelectro-elasticity and megneto-elasticity as well
as elastic wave phenomena. In this work, Kazumi Tanuma presents the essential elements of
the Stroh formalism, its major theorems with proofs and applications, in a research-level
textbook form. The presentation is self-contained and the development is clear and concise.
Professor Tanuma's efforts to produce a straightforward, accurate and readable account of
this subject are clearly evident. He has organized the subject with a common logical thread
throughout and he has given basis for not only the importance of this area of research, but,
most importantly, for understanding its meaning and significance.

R. Fosdick (✉)
Department of Aerospace Engineering and Mechanics, University of Minnesota,
110 Union St SE, Minneapolis, MN 55455, USA
e-mail: fosdick@aem.umn.edu

Foreword. R. Fosdick (ed.)
doi: 10.1007/978-1-4020-6388-6_1, © Springer Science + Business Media B.V. 2007

Preface

Kazumi Tanuma

The Stroh formalism is a powerful and elegant mathematical method developed for the analysis of the equations of anisotropic elasticity. The purpose of this exposition is to introduce the essence of this formalism and demonstrate its effectiveness in both static and dynamic elasticity. The Stroh formalism is useful for constructing the solutions to the equations of anisotropic elasticity and for investigating the properties which these solutions possess. The equations of elasticity are complicated, because they constitute a system and, particularly for the anisotropic cases, inherit many parameters from the elasticity tensor. The Stroh formalism reveals simple structures hidden in the equations of anisotropic elasticity and provides a systematic approach to these equations. This fact is not only interesting in itself from a mathematical point of view, but also is important in that the Stroh formalism helps us compute physical quantities explicitly.

This exposition is divided into three chapters. Chapter 1 gives a succinct introduction to the Stroh formalism so that the reader could grasp the essentials as quickly as possible. A basic element of the Stroh formalism is Stroh's eigenvalue problem, each of whose six-dimensional eigenvectors and generalized eigenvectors is composed of a three-dimensional displacement part and a three-dimensional traction part that pertain to two-dimensional deformations in three-dimensional elasticity. The most fundamental parts of this chapter are Section 1.3 and Section 1.5, where rotational invariance and rotational dependence of the Stroh eigenvectors and generalized eigenvectors in the reference plane are proved. On the basis of these results, the Barnett-Lothe integral formalism is developed and the notion of the surface impedance tensor is introduced.

In Chapter 2 several important topics in static elasticity, which include fundamental solutions, piezoelectricity, and inverse boundary value problems, are studied on the basis of the Stroh formalism. The surface impedance tensor plays important

K. Tanuma (✉)
Department of Mathematics, Graduate School of Engineering,
Gunma University, Kiryu 376-8515, Japan
e-mail: tanuma@math.sci.gunma-u.ac.jp

Preface. K.Tanuma
doi: 10.1007/978-1-4020-6388-6_2, © Springer Science + Business Media B.V. 2007

roles in these studies. The Stroh formalism was developed for two-dimensional deformations in three-dimensional elasticity. But the formalism can also be useful in full three-dimensional problems, as we shall see in this chapter.

Chapter 3 is devoted to Rayleigh waves, for long a topic of utmost importance in nondestructive evaluation, seismology, and materials science. There we discuss existence, uniqueness, phase velocity, polarization, and perturbation of the phase velocity of Rayleigh waves through the Stroh formalism. We put emphasis on the usefulness of $S_3(v)$, which is one of the three real matrices introduced in the Barnett-Lothe integral formalism. For instance, the phase velocity of Rayleigh waves can be analyzed by examining only one diagonal component of $S_3(v)$.

Chapter 1 is the foundation for both Chapter 2 and Chapter 3. Particularly, quite a number of results in the Stroh formalism for static elasticity apply in a parallel way to dynamic elasticity. Hence even readers who are primarily interested in applications of the Stroh formalism to Rayleigh waves are urged to read Chapter 1 before they proceed to Chapter 3.

We give the subjects described above as self-contained a treatment as possible, although we do refer the reader to the literature for the long proofs of several lemmas. We also present practical examples of algebraic calculations as clearly as possible, which will make the Stroh formalism more understandable and attractive.

The first draft of several basic sections of this exposition was written as lecture notes for a series of talks I gave, on the invitation of Chi-Sing Man, to graduate students at the University of Kentucky in the spring of 2001. Since then, I have been able to add many subjects, which include joint works with Chi-Sing Man, Gen Nakamura and Masayuki Akamatsu.

I owe most to Chi-Sing Man, whose kind suggestion two years ago prompted me to write this exposition. He carefully read the complete manuscript, corrected sentences, and gave me many valuable and important comments and criticism, which not only highly improved this exposition but also let myself learn much on elasticity. Without his help I could not have completed this exposition. I am indebted to Roger Fosdick, who gave me valuable suggestions, constant support, and the chance for this exposition to be published. I would also like to express my thanks to Gen Nakamura, whose valuable guidance has led me to study the equations of elasticity and to learn the Stroh formalism. My thanks go also to Takao Kakita, my adviser, who introduced me to the study on partial differential equations when I was a graduate student. He always gave me valuable suggestions and encouragement. I would also like to thank Yuusuke Iso, who has always shown much interest in my work and has given many helpful suggestions. I am also grateful to my friends and colleagues Masayuki Akamatsu, Ryuichi Ashino, Hiroya Ito, and Mishio Kawashita for valuable discussions with them and helpful advice from them. Lastly I thank Nathalie Jacobs, Anneke Pot, Fiona Routley and Claudia Magpantay of Springer for their kind correspondences with me these two years. This work was partly supported by Grant-in-Aid for Scientific Research (C) (Nos. 16540095 & 19540113), Society for the Promotion of Science, Japan.

Kiryu, Japan, March 2007

Stroh Formalism and Rayleigh Waves

Kazumi Tanuma

Abstract The Stroh formalism is a powerful and elegant mathematical method developed for the analysis of the equations of anisotropic elasticity. The purpose of this exposition is to introduce the essence of this formalism and demonstrate its effectiveness in both static and dynamic elasticity. The equations of elasticity are complicated, because they constitute a system and, particularly for the anisotropic cases, inherit many parameters from the elasticity tensor. The Stroh formalism reveals simple structures hidden in the equations of anisotropic elasticity and provides a systematic approach to these equations. This exposition is divided into three chapters. Chapter 1 gives a succinct introduction to the Stroh formalism so that the reader could grasp the essentials as quickly as possible. In Chapter 2 several important topics in static elasticity, which include fundamental solutions, piezoelectricity, and inverse boundary value problems, are studied on the basis of the Stroh formalism. Chapter 3 is devoted to Rayleigh waves, for long a topic of utmost importance in nondestructive evaluation, seismology, and materials science. There we discuss existence, uniqueness, phase velocity, polarization, and perturbation of Rayleigh waves through the Stroh formalism.

Keywords Anisotropic elasticity · Rayleigh waves · The Stroh formalism · Equations of elasticity · Inverse problems

Mathematics Subject Classifications (2000) 74B05 · 74E10 · 74G75 · 74J15

The Table of Contents and Index are also provided as Electronic Supplementary Material for online readers at doi:10.1007/s10659-007-9117-1

K. Tanuma (✉)
Department of Mathematics, Graduate School of Engineering,
Gunma University, Kiryu 376-8515, Japan
e-mail: tanuma@math.sci.gunma-u.ac.jp

Stroh Formalism and Rayleigh Waves. K.Tanuma
doi: 10.1007/978-1-4020-6388-6_3, © Springer Science + Business Media B.V. 2007

1 The Stroh Formalism for Static Elasticity

1.1 Basic Elasticity

Let a Cartesian coordinate system be chosen in space, and let $\mathbb{R}$ denote the set of real numbers. Suppose that a region B in $\mathbb{R}^3$ with a smooth boundary is occupied by an elastic medium. Let S be a regular oriented surface in the region B and let $\mathbf{n} = (n_1, n_2, n_3)$ be the unit positive normal to S at $\mathbf{x} \in B$.

The stress tensor $\boldsymbol{\sigma} = (\sigma_{ij})_{i,j=1,2,3}$ at $\mathbf{x}$ is defined so that the vector

$$\left(\sum_{j=1}^{3} \sigma_{ij} n_j \right)_{i\downarrow 1,2,3} \tag{1.1}$$

represents the traction or force per unit area at $\mathbf{x}$ exerted by the portion of B on the side of S toward which $\mathbf{n}$ points on the portion of B on the other side.

Let $\boldsymbol{\varepsilon} = \boldsymbol{\varepsilon}(\boldsymbol{u}) = (\varepsilon_{ij})_{i,j=1,2,3}$ be the infinitesimal strain tensor

$$\varepsilon_{ij} = \frac{1}{2} \left(\frac{\partial u_i}{\partial x_j} + \frac{\partial u_j}{\partial x_i} \right), \tag{1.2}$$

where $\boldsymbol{u} = \boldsymbol{u}(\mathbf{x}) = (u_1, u_2, u_3)$ is the displacement at the place $\mathbf{x}$ and (x_1, x_2, x_3) are the Cartesian coordinates of $\mathbf{x}$.

In linearized elasticity, the stresses σ_{ij} are related to the strains ε_{ij} by the generalized Hooke's law

$$\sigma_{ij} = \sum_{k,l=1}^{3} C_{ijkl}\, \varepsilon_{kl}, \qquad i, j = 1, 2, 3. \tag{1.3}$$

Here $\mathbf{C} = \mathbf{C}(\mathbf{x}) = (C_{ijkl})_{i,j,k,l=1,2,3}$ is the elasticity tensor. Since the law of balance of angular momentum implies that

$$\sigma_{ij} = \sigma_{ji}$$

and (1.2) implies that $\varepsilon_{ij} = \varepsilon_{ji}$, we may assume that $\mathbf{C}$ has the minor symmetries

$$C_{ijkl} = C_{jikl}, \qquad C_{ijkl} = C_{ijlk}, \qquad i, j, k, l = 1, 2, 3. \tag{1.4}$$

Then (1.3) can be written as

$$\sigma_{ij} = \sum_{k,l=1}^{3} C_{ijkl} \frac{\partial u_k}{\partial x_l}, \qquad i, j = 1, 2, 3. \tag{1.5}$$

We also assume the existence of a stored energy function $W = W(\boldsymbol{\varepsilon})$ such that

$$W(\mathbf{0}) = 0, \qquad \frac{\partial}{\partial \varepsilon_{ij}} W = \sigma_{ij},$$

 Springer

which, combined with (1.3), implies that

$$\frac{\partial}{\partial \varepsilon_{kl}}\left(\frac{\partial}{\partial \varepsilon_{ij}}W\right) = C_{ijkl}, \qquad \frac{\partial}{\partial \varepsilon_{ij}}\left(\frac{\partial}{\partial \varepsilon_{kl}}W\right) = C_{klij}.$$

Hence $\mathbf{C}$ has the major symmetry

$$C_{ijkl} = C_{klij}, \qquad i, j, k, l = 1, 2, 3. \tag{1.6}$$

We assume that the stored energy function is positive for any non-zero strain

$$W(\boldsymbol{\varepsilon}) > 0 \text{ for } \boldsymbol{\varepsilon} \neq \mathbf{0},$$

which implies that $\mathbf{C}$ is positive definite or it satisfies the following strong convexity condition:

$$\sum_{i,j,k,l=1}^{3} C_{ijkl}\, \varepsilon_{ij}\, \varepsilon_{kl} > 0 \quad \text{for any non-zero } 3 \times 3 \text{ real symmetric matrix } (\varepsilon_{ij}). \tag{1.7}$$

In Chapter 1 and Chapter 2 we study static deformations of the elastic medium in B with no body force. Then the displacement $\boldsymbol{u}$ is independent of time and the equations of equilibrium are written as

$$\sum_{j=1}^{3} \frac{\partial}{\partial x_j}\sigma_{ij} = 0, \qquad i = 1, 2, 3. \tag{1.8}$$

By (1.5), the equations of equilibrium can be expressed in terms of the displacement $\mathbf{u}$ as

$$\sum_{j,k,l=1}^{3} \frac{\partial}{\partial x_j}\left(C_{ijkl}\frac{\partial u_k}{\partial x_l}\right) = 0, \qquad i = 1, 2, 3. \tag{1.9}$$

The elasticity tensor $\mathbf{C} = \mathbf{C}(\mathbf{x}) = \left(C_{ijkl}\right)_{i,j,k,l=1,2,3}$ appearing in (1.3) determines the elastic response of the material point in question. From the symmetries (1.4) and (1.6) it follows that $\mathbf{C} = \mathbf{C}(\mathbf{x})$ has 21 independent components at each $\mathbf{x}$.

We say that the elastic material at $\mathbf{x}$ is isotropic if the elasticity tensor $\mathbf{C}(\mathbf{x}) = \left(C_{ijkl}\right)_{i,j,k,l=1,2,3}$ satisfies

$$C_{ijkl} = \sum_{p,q,r,s=1}^{3} Q_{ip}\, Q_{jq}\, Q_{kr}\, Q_{ls}\, C_{pqrs}, \qquad i, j, k, l = 1, 2, 3 \tag{1.10}$$

Springer

for any orthogonal tensor $Q = (Q_{ij})_{i,j=1,2,3}$.[1] For isotropic materials, the components C_{ijkl} $(i, j, k, l = 1, 2, 3)$ are determined by two independent parameters and are given by the representation formula

$$C_{ijkl} = \lambda \, \delta_{ij}\delta_{kl} + \mu(\delta_{ik}\delta_{jl} + \delta_{il}\delta_{kj}), \qquad i, j, k, l = 1, 2, 3, \qquad (1.11)$$

where δ_{ij} is Kronecker's delta symbol and $\lambda = \lambda(\mathbf{x})$, $\mu = \mu(\mathbf{x})$ are scalars called the Lamé moduli.

We say that the elastic material at $\mathbf{x}$ is anisotropic if it is not isotropic. The largest proper subgroup $\mathcal{G}$ of the orthogonal group such that (1.10) holds for any $Q = (Q_{ij})_{i,j=1,2,3}$ in $\mathcal{G}$ is called the symmetry group of the anisotropic material at $\mathbf{x}$.

According to the generators of the subgroup $\mathcal{G}$, the number of independent components in $\mathbf{C}(\mathbf{x}) = (C_{ijkl})_{i,j,k,l=1,2,3}$ varies and anisotropic elastic materials can be classified (see, for example, Sections 21 and 26 of [33]). To express the components C_{ijkl} $(i, j, k, l = 1, 2, 3)$ conveniently, we introduce the contracted notation called the Voigt notation:

We rewrite $(C_{ijkl})_{i,j,k,l=1,2,3}$ using the 6×6 matrix as

$$(C_{\alpha\beta}) = \begin{bmatrix} C_{11} & C_{12} & C_{13} & C_{14} & C_{15} & C_{16} \\ & C_{22} & C_{23} & C_{24} & C_{25} & C_{26} \\ & & C_{33} & C_{34} & C_{35} & C_{36} \\ & & & C_{44} & C_{45} & C_{46} \\ & & & & C_{55} & C_{56} \\ \text{Sym.} & & & & & C_{66} \end{bmatrix}, \qquad (1.12)$$

where we have used the rules for replacing the subscript ij (or kl) by α (or β) as follows:

ij (or kl)	α or β	ij (or kl)	α or β
11	$\longleftrightarrow$ 1	23 or 32	$\longleftrightarrow$ 4
22	$\longleftrightarrow$ 2	31 or 13	$\longleftrightarrow$ 5
33	$\longleftrightarrow$ 3	12 or 21	$\longleftrightarrow$ 6.

This replacement is possible because of the minor symmetries (1.4). The symmetry of the matrix (1.12) follows from the major symmetry (1.6). Note that the strong convexity condition (1.7) is equivalent to the assertion that the 6×6 matrix (1.12) is positive definite (Exercise 1-1).

Now let $R_{x_i}^{\phi}$ denote the orthogonal tensor corresponding to a right-handed rotation by the angle ϕ about the i-axis of the Cartesian coordinate system. We give here several typical examples of anisotropic elastic materials with the generators of

---

[1] If we write the generalized Hooke's law (1.3) as $\sigma = \sigma(\varepsilon)$, it can be proved that (1.10) is equivalent to

$$Q\sigma(\varepsilon)\, Q^T = \sigma(Q\varepsilon\, Q^T),$$

which implies that the transformed σ under the orthogonal tensor Q is related to the transformed ε under the orthogonal tensor Q by the same Hooke's law. Here σ and ε are interpreted as the 3×3 matrices whose (i, j) components are σ_{ij} and ε_{ij} respectively, and the superscript T denotes transposition.

their symmetry groups and write the corresponding elasticity tensors in the Voigt notations by using their independent components.

(1) Transversely isotropic materials.
   The generators of $\mathcal{G}$ are $\boldsymbol{R}_{x_3}^{\phi}, 0 < \phi < 2\pi,$[2] and then

$$(C_{\alpha\beta}) = \begin{bmatrix} C_{11} & C_{12} & C_{13} & 0 & 0 & 0 \\ & C_{11} & C_{13} & 0 & 0 & 0 \\ & & C_{33} & 0 & 0 & 0 \\ & & & C_{44} & 0 & 0 \\ & & & & C_{44} & 0 \\ \text{Sym.} & & & & & \frac{C_{11}-C_{12}}{2} \end{bmatrix}. \tag{1.13}$$

(2) Cubic materials.
   The generators of $\mathcal{G}$ are $\boldsymbol{R}_{x_1}^{\frac{\pi}{2}}, \boldsymbol{R}_{x_2}^{\frac{\pi}{2}}, \boldsymbol{R}_{x_3}^{\frac{\pi}{2}},$ and then

$$(C_{\alpha\beta}) = \begin{bmatrix} C_{11} & C_{12} & C_{12} & 0 & 0 & 0 \\ & C_{11} & C_{12} & 0 & 0 & 0 \\ & & C_{11} & 0 & 0 & 0 \\ & & & C_{44} & 0 & 0 \\ & & & & C_{44} & 0 \\ \text{Sym.} & & & & & C_{44} \end{bmatrix}. \tag{1.14}$$

(3) Orthorhombic materials.
   The generators of $\mathcal{G}$ are $\boldsymbol{R}_{x_2}^{\pi}, \boldsymbol{R}_{x_3}^{\pi},$ and then

$$(C_{\alpha\beta}) = \begin{bmatrix} C_{11} & C_{12} & C_{13} & 0 & 0 & 0 \\ & C_{22} & C_{23} & 0 & 0 & 0 \\ & & C_{33} & 0 & 0 & 0 \\ & & & C_{44} & 0 & 0 \\ & & & & C_{55} & 0 \\ \text{Sym.} & & & & & C_{66} \end{bmatrix}. \tag{1.15}$$

(4) Monoclinic materials.
   The generator of $\mathcal{G}$ is $\boldsymbol{R}_{x_3}^{\pi},$ and then

$$(C_{\alpha\beta}) = \begin{bmatrix} C_{11} & C_{12} & C_{13} & 0 & 0 & C_{16} \\ & C_{22} & C_{23} & 0 & 0 & C_{26} \\ & & C_{33} & 0 & 0 & C_{36} \\ & & & C_{44} & C_{45} & 0 \\ & & & & C_{55} & 0 \\ \text{Sym.} & & & & & C_{66} \end{bmatrix}. \tag{1.16}$$

Note that all these symmetries are, in principle, defined at each point $\mathbf{x}$.

When $\boldsymbol{R}_{x_i}^{\theta}$ belongs to the symmetry group $\mathcal{G}$ of the material at $\mathbf{x}$, we say that the i-axis is a $\frac{2\pi}{\theta}$-fold symmetry axis at $\mathbf{x}$. For example, for cubic materials described above, the 3-axis is a 4-fold symmetry axis.

---

[2]In this case the 3-axis is called the axis of symmetry. For hexagonal materials, the generator of $\mathcal{G}$ are $\boldsymbol{R}_{x_2}^{\pi}$ and $\boldsymbol{R}_{x_3}^{\frac{\pi}{3}}$. It can be proved that hexagonal materials and transversely isotropic materials have elasticity tensors of the same form.

Generally, the elasticity tensor $\mathbf{C}$ depends on $\mathbf{x}$. However, if $\mathbf{C}$ is independent of $\mathbf{x}$, we say that the elastic body is homogeneous. In this case, (1.9) becomes

$$\sum_{j,k,l=1}^{3} C_{ijkl} \frac{\partial^2 u_k}{\partial x_j \partial x_l} = 0, \qquad i = 1, 2, 3.$$

In the subsequent sections of this chapter we develop the Stroh formalism, where the elastic body in question is assumed to be homogeneous.

1.2 Stroh's Eigenvalue Problem

Let $\mathbf{x} = (x_1, x_2, x_3)$ be the position vector and let $\mathbf{m} = (m_1, m_2, m_3)$ and $\mathbf{n} = (n_1, n_2, n_3)$ be orthogonal unit vectors in $\mathbb{R}^3$. Let $\mathbb{C}$ denote the set of complex numbers.

We consider the equations of equilibrium for the homogeneous elastic body in static deformations in the half-space $\mathbf{n} \cdot \mathbf{x} = n_1 x_1 + n_2 x_2 + n_3 x_3 \leq 0$;[3]

$$\sum_{j,k,l=1}^{3} C_{ijkl} \frac{\partial^2 u_k}{\partial x_j \partial x_l} = 0, \qquad i = 1, 2, 3. \tag{1.17}$$

Here $\mathbf{C} = \left(C_{ijkl}\right)_{i,j,k,l=1,2,3}$ is the elasticity tensor, which has the major symmetry (1.6) and the minor symmetries (1.4) and satisfies the strong convexity condition (1.7). We seek solutions to (1.17) of the form

$$\boldsymbol{u} = (u_1, u_2, u_3) = \mathbf{a}\, e^{-\sqrt{-1}\xi(\mathbf{m}\cdot\mathbf{x} + p\,\mathbf{n}\cdot\mathbf{x})} \in \mathbb{C}^3 \tag{1.18}$$

in the half-space $\mathbf{n} \cdot \mathbf{x} \leq 0$, each of which satisfies the physical requirement that the amplitude decays with depth below the surface $\mathbf{n} \cdot \mathbf{x} = 0$; here $\mathbf{a} \in \mathbb{C}^3$, $p \in \mathbb{C}$, and ξ is a real parameter that is positive.[4]

A solution of the form (1.18) describes two-dimensional deformations in an elastic body; it depends on the projections of $\mathbf{x}$ on the plane spanned by the two orthogonal unit vectors $\mathbf{m}$ and $\mathbf{n}$. Such a plane is called the reference plane.

Let us determine $\mathbf{a} \in \mathbb{C}^3$ and $p \in \mathbb{C}$ in (1.18). Substituting (1.18) into (1.17) and noting that

$$\frac{\partial \boldsymbol{u}}{\partial x_j} = -\sqrt{-1}(m_j + p n_j)\,\xi\,\mathbf{a}\, e^{-\sqrt{-1}\xi(\mathbf{m}\cdot\mathbf{x} + p\,\mathbf{n}\cdot\mathbf{x})},$$

---

[3]Throughout this exposition, the dot ' $\cdot$ ' denotes the inner product of two vectors in $\mathbb{R}^3$. When denoting the inner product of two vectors in $\mathbb{C}^3$, we use the notation $(\ ,\ )_{\mathbb{C}^3}$.

[4]The positive parameter ξ carries the dimension of the reciprocal of length. The rationale for studying this special class of solutions will be given in Section 1.9. See also Remark 1.8.

we get

$$\left(\sum_{j,l=1}^{3} C_{ijkl}\, (m_j + pn_j)(m_l + pn_l)\right)_{i\downarrow k\rightarrow 1,2,3} \mathbf{a}$$

$$= \left(\sum_{j,l=1}^{3} C_{ijkl}\, m_j m_l + p\left(\sum_{j,l=1}^{3} C_{ijkl}\, m_j n_l + \sum_{j,l=1}^{3} C_{ijkl}\, n_j m_l\right)\right.$$

$$\left. + p^2 \sum_{j,l=1}^{3} C_{ijkl}\, n_j n_l\right)_{i\downarrow k\rightarrow 1,2,3} \mathbf{a} = \mathbf{0}. \qquad (1.19)$$

We introduce 3×3 real matrices

$$\mathbf{Q} = \left(\sum_{j,l=1}^{3} C_{ijkl}\, m_j m_l\right)_{i\downarrow k\rightarrow 1,2,3}, \quad \mathbf{R} = \left(\sum_{j,l=1}^{3} C_{ijkl}\, m_j n_l\right)_{i\downarrow k\rightarrow 1,2,3},$$

$$\mathbf{T} = \left(\sum_{j,l=1}^{3} C_{ijkl}\, n_j n_l\right)_{i\downarrow k\rightarrow 1,2,3}. \qquad (1.20)$$

Then it follows from the major symmetry (1.6) that

$$\mathbf{R}^T = \left(\sum_{j,l=1}^{3} C_{ijkl}\, n_j m_l\right)_{i\downarrow k\rightarrow 1,2,3},$$

and hence equation (1.19) can be written as

$$[\mathbf{Q} + p(\mathbf{R} + \mathbf{R}^T) + p^2\mathbf{T}]\,\mathbf{a} = \mathbf{0}, \qquad (1.21)$$

where the superscript T denotes transposition. For the existence of a non-trivial vector $\mathbf{a} \neq \mathbf{0}$, we observe that p satisfies the sextic equation

$$\det\left[\mathbf{Q} + p(\mathbf{R} + \mathbf{R}^T) + p^2\mathbf{T}\right] = 0. \qquad (1.22)$$

For the matrices $\mathbf{Q}$, $\mathbf{T}$ and the solutions p to (1.22), we have

Lemma 1.1

(1) *The matrices $\mathbf{Q}$ and $\mathbf{T}$ are symmetric and positive definite.*
(2) *The characteristic roots p_α ($1 \leq \alpha \leq 6$), i.e., the solutions to the sextic equation (1.22), are not real and they occur in complex conjugate pairs.*

Proof That the matrices $\mathbf{Q}$ and $\mathbf{T}$ are symmetric follows from definition (1.20) and the major symmetry (1.6). For any non-zero $\mathbf{v} = (v_1, v_2, v_3)^T \in \mathbb{R}^3$, put

$$\varepsilon_{ij} = m_i v_j + v_i m_j.$$

🖂 Springer

Then (ε_{ij}) is a non-zero symmetric matrix. The strong convexity condition (1.7), combined with the minor symmetries (1.4), leads to

$$0 < \sum_{i,j,k,l=1}^{3} C_{ijkl}\,\varepsilon_{ij}\,\varepsilon_{kl} = 4 \sum_{i,k=1}^{3} v_i \left(\sum_{j,l=1}^{3} C_{ijkl}\,m_j m_l \right) v_k = 4(\mathbf{v} \cdot \mathbf{Q}\,\mathbf{v}),$$

which implies that $\mathbf{Q}$ is positive definite. In the same way, it is proved that $\mathbf{T}$ is positive definite.

Next we prove (2). The solutions p_α $(1 \leq \alpha \leq 6)$ to (1.22) are not real if and only if the matrix

$$\left(\sum_{j,l=1}^{3} C_{ijkl}\,(m_j + pn_j)(m_l + pn_l) \right)_{i \downarrow k \to 1,2,3} \qquad \text{is invertible for all } p \in \mathbb{R}. \qquad (1.23)$$

Since the vector

$$\mathbf{w} = (w_1, w_2, w_3) = \frac{\mathbf{m} + p\mathbf{n}}{|\mathbf{m} + p\mathbf{n}|} \qquad (p \in \mathbb{R})$$

moves on the unit circle, it follows from (1) of the lemma that the symmetric matrix

$$\left(\sum_{j,l=1}^{3} C_{ijkl}\,w_j w_l \right)_{i \downarrow k \to 1,2,3}$$

is positive definite, which implies (1.23). Since all the coefficients of the sextic equation (1.22) are real, its roots occur in complex conjugate pairs. □

Next we give the traction on the surface $\mathbf{n} \cdot \mathbf{x} = 0$ produced by the displacement (1.18). Since the unit outward normal of this surface is the vector $\mathbf{n}$, the traction t on $\mathbf{n} \cdot \mathbf{x} = 0$ is given by

$$t = \left(\sum_{j=1}^{3} \sigma_{ij}\,n_j \right)_{i \downarrow 1,2,3} \Big|_{\mathbf{n} \cdot \mathbf{x}=0}$$

which, by (1.5), becomes

$$t = \left(\sum_{j,k,l=1}^{3} C_{ijkl}\,\frac{\partial u_k}{\partial x_l}\,n_j \right)_{i \downarrow 1,2,3} \Big|_{\mathbf{n} \cdot \mathbf{x}=0}. \qquad (1.24)$$

It follows from (1.18) and (1.20) that

$$t = -\sqrt{-1}\,\xi \left(\sum_{j,l=1}^{3} C_{ijkl}\,(m_l + pn_l)\,n_j \right)_{i \downarrow k \to 1,2,3} \mathbf{a}\,e^{-\sqrt{-1}\,\xi\,\mathbf{m} \cdot \mathbf{x}}$$

$$= -\sqrt{-1}\,\xi\,\left[\mathbf{R}^T + p\mathbf{T} \right]\,\mathbf{a}\,e^{-\sqrt{-1}\,\xi\,\mathbf{m} \cdot \mathbf{x}}.$$

Hence we define a vector $\mathbf{l} \in \mathbb{C}^3$ as

$$\mathbf{l} = \left[\mathbf{R}^T + p\mathbf{T} \right]\,\mathbf{a}. \qquad (1.25)$$

Then

$$t = -\sqrt{-1}\,\xi\,\mathbf{l}\,e^{-\sqrt{-1}\,\xi\,\mathbf{m}\cdot\mathbf{x}} \tag{1.26}$$

is the traction on the surface $\mathbf{n}\cdot\mathbf{x} = 0$.

Finally we shall see that the relations (1.21) and (1.25) can be recast as a six-dimensional eigenvalue problem, called Stroh's eigenvalue problem.

By (1) of Lemma 1.1, $\mathbf{T}^{-1}$ exists. Hence from (1.25) we get

$$p\,\mathbf{a} = -\mathbf{T}^{-1}\mathbf{R}^{T}\,\mathbf{a} + \mathbf{T}^{-1}\mathbf{l} \tag{1.27}$$

and

$$p\,\mathbf{l} = \left[p\,\mathbf{R}^{T} + p^{2}\mathbf{T}\right]\mathbf{a}.$$

The last equation becomes, by (1.21),

$$p\,\mathbf{l} = -[\mathbf{Q} + p\,\mathbf{R}]\,\mathbf{a},$$

and from (1.27)

$$p\,\mathbf{l} = -\mathbf{Q}\,\mathbf{a} - \mathbf{R}\left(-\mathbf{T}^{-1}\mathbf{R}^{T}\,\mathbf{a} + \mathbf{T}^{-1}\,\mathbf{l}\right) = \left[-\mathbf{Q} + \mathbf{R}\mathbf{T}^{-1}\mathbf{R}^{T}\right]\mathbf{a} - \mathbf{R}\mathbf{T}^{-1}\mathbf{l}. \tag{1.28}$$

Thus, from (1.27) and (1.28) we obtain

Theorem 1.2 *Let* $\begin{bmatrix}\mathbf{a}\\\mathbf{l}\end{bmatrix}$ *be a column vector in* $\mathbb{C}^{6}$ *whose first three components consist of a vector* $\mathbf{a} \in \mathbb{C}^{3}$ *that satisfies (1.21) and whose last three components consist of the vector* $\mathbf{l} \in \mathbb{C}^{3}$ *given by (1.25). Then the following six-dimensional eigenrelation holds:*

$$\mathbf{N}\begin{bmatrix}\mathbf{a}\\\mathbf{l}\end{bmatrix} = p\begin{bmatrix}\mathbf{a}\\\mathbf{l}\end{bmatrix}, \tag{1.29}$$

where $\mathbf{N}$ *is the* 6×6 *real matrix defined by*

$$\mathbf{N} = \begin{bmatrix} -\mathbf{T}^{-1}\mathbf{R}^{T} & \mathbf{T}^{-1} \\ -\mathbf{Q} + \mathbf{R}\mathbf{T}^{-1}\mathbf{R}^{T} & -\mathbf{R}\mathbf{T}^{-1} \end{bmatrix}. \tag{1.30}$$

A solution $\begin{bmatrix}\mathbf{a}\\\mathbf{l}\end{bmatrix}$ to (1.29) is called a Stroh eigenvector, where $\mathbf{a}$ and $\mathbf{l}$ are its displacement part and traction part, respectively. Lemma 1.1 implies that the eigenvalues p_{α} ($1 \le \alpha \le 6$) of $\mathbf{N}$ are not real and they occur in complex conjugate pairs. We call the eigenvalue problem (1.29) Stroh's eigenvalue problem.

As we shall see, coupling the displacement part and the traction part together in (1.29) leads us to the finding of an elegant result on rotational dependency of these vectors in the reference plane, which could not be proved if we restrict ourselves to consideration of displacements (i.e., the solutions to (1.17)).

1.3 Rotational Invariance of Stroh Eigenvector in Reference Plane

The displacement part $\mathbf{a} \in \mathbb{C}^3$ and the traction part $\mathbf{l} \in \mathbb{C}^3$ of a Stroh eigenvector are determined by (1.21) and (1.25). These vectors $\mathbf{a}$ and $\mathbf{l}$ (and also the characteristic roots p which are the solutions to (1.22)) apparently depend on the two orthogonal unit vectors $\mathbf{m}$ and $\mathbf{n}$. In this section we show that $\mathbf{a}$ and $\mathbf{l}$ do not depend on $\mathbf{m}$ and $\mathbf{n}$ individually but depend on the vector product $\mathbf{m} \times \mathbf{n}$. This means that $\mathbf{a}$ and $\mathbf{l}$ are invariant under the rotations of $\mathbf{m}$ and $\mathbf{n}$ around $\mathbf{m} \times \mathbf{n}$, and hence are rotationally invariant in the reference plane. This property is fundamental in the Stroh formalism and is also elegant in itself.

We formulate the above more precisely. Let us fix two orthogonal unit vectors $\mathbf{e}_1$ and $\mathbf{e}_2$ in $\mathbb{R}^3$ and take $\mathbf{m}$ and $\mathbf{n}$ to be those which are obtained by rotating $\mathbf{e}_1$ and $\mathbf{e}_2$ around $\mathbf{e}_1 \times \mathbf{e}_2$ by an angle ϕ ($0 \le \phi < 2\pi$) so that

$$\mathbf{m} = \mathbf{m}(\phi) = \mathbf{e}_1 \cos \phi + \mathbf{e}_2 \sin \phi, \qquad \mathbf{n} = \mathbf{n}(\phi) = -\mathbf{e}_1 \sin \phi + \mathbf{e}_2 \cos \phi. \tag{1.31}$$

Then $\mathbf{Q}, \mathbf{R}, \mathbf{T}$ in (1.20), and hence p in (1.22), can be regarded as functions of ϕ

$$\mathbf{Q} = \mathbf{Q}(\phi), \quad \mathbf{R} = \mathbf{R}(\phi), \quad \mathbf{T} = \mathbf{T}(\phi), \quad p = p(\phi) \tag{1.32}$$

and by (1.30),

$$\mathbf{N} = \mathbf{N}(\phi) = \begin{bmatrix} -\mathbf{T}(\phi)^{-1}\mathbf{R}(\phi)^T & \mathbf{T}(\phi)^{-1} \\ -\mathbf{Q}(\phi) + \mathbf{R}(\phi)\mathbf{T}(\phi)^{-1}\mathbf{R}(\phi)^T & -\mathbf{R}(\phi)\mathbf{T}(\phi)^{-1} \end{bmatrix}. \tag{1.33}$$

Theorem 1.3 *Let $\mathbf{a}$ and $\mathbf{l}$ satisfy (1.21) and (1.25) at $\phi = 0$ with $p = p_1$ being a solution to (1.22) at $\phi = 0$. Then the same $\mathbf{a}$ and $\mathbf{l}$ satisfy (1.21) and (1.25) for all ϕ, while $p = p(\phi)$ is the solution to the Riccati equation*

$$\frac{d}{d\phi} p = -1 - p^2 \tag{1.34}$$

with $p(0) = p_1$.

Proof We denote the differentiation $\frac{d}{d\phi}$ by $'$. From (1.31) we get

$$\mathbf{m}' = -\mathbf{e}_1 \sin \phi + \mathbf{e}_2 \cos \phi = \mathbf{n}, \qquad \mathbf{n}' = -\mathbf{e}_1 \cos \phi - \mathbf{e}_2 \sin \phi = -\mathbf{m}.$$

 Springer

Then from (1.20) it follows that

$$
\begin{aligned}
\mathbf{Q}' &= \left(\sum_{j,l=1}^{3} C_{ijkl}\, m_j m_l \right)'_{i\downarrow k \to 1,2,3} \\
&= \left(\sum_{j,l=1}^{3} C_{ijkl}\, n_j m_l + \sum_{j,l=1}^{3} C_{ijkl}\, m_j n_l \right)_{i\downarrow k \to 1,2,3} = \mathbf{R} + \mathbf{R}^T, \\
\mathbf{R}' &= \left(\sum_{j,l=1}^{3} C_{ijkl}\, m_j n_l \right)'_{i\downarrow k \to 1,2,3} \\
&= \left(\sum_{j,l=1}^{3} C_{ijkl}\, n_j n_l - \sum_{j,l=1}^{3} C_{ijkl}\, m_j m_l \right)_{i\downarrow k \to 1,2,3} = \mathbf{T} - \mathbf{Q} \\
&= (\mathbf{R}^T)', \\
\mathbf{T}' &= \left(\sum_{j,l=1}^{3} C_{ijkl}\, n_j n_l \right)'_{i\downarrow k \to 1,2,3} \\
&= -\left(\sum_{j,l=1}^{3} C_{ijkl}\, m_j n_l + \sum_{j,l=1}^{3} C_{ijkl}\, n_j m_l \right)_{i\downarrow k \to 1,2,3} = -\left(\mathbf{R} + \mathbf{R}^T \right). \quad (1.35)
\end{aligned}
$$

Put

$$
\mathbf{H}(\phi) = \left[\mathbf{Q} + p\left(\mathbf{R} + \mathbf{R}^T \right) + p^2 \mathbf{T} \right] (\phi).
$$

From (1.35) we have

$$
\begin{aligned}
\mathbf{H}'(\phi) &= \left[\mathbf{R} + \mathbf{R}^T + p'\left(\mathbf{R} + \mathbf{R}^T \right) + 2p\,(\mathbf{T} - \mathbf{Q}) + 2p\,p'\,\mathbf{T} - p^2 \left(\mathbf{R} + \mathbf{R}^T \right) \right] (\phi) \\
&= \left[-2p\,\mathbf{Q} + \left(1 + p' - p^2 \right) \left(\mathbf{R} + \mathbf{R}^T \right) + 2p\left(1 + p' \right) \mathbf{T} \right] (\phi).
\end{aligned}
$$

Suppose that $p(\phi)$ satisfies the Riccati equation (1.34). Then the equation above becomes

$$
\mathbf{H}'(\phi) = -2p\left[\mathbf{Q} + p(\mathbf{R} + \mathbf{R}^T) + p^2 \mathbf{T} \right] (\phi) = -2p(\phi)\,\mathbf{H}(\phi). \quad (1.36)
$$

Now put

$$
\mathbf{h}(\phi) = \left[\mathbf{Q} + p(\mathbf{R} + \mathbf{R}^T) + p^2 \mathbf{T} \right] (\phi)\, \mathbf{a} = \mathbf{H}(\phi)\, \mathbf{a},
$$

where $\mathbf{a}$ satisfies (1.21) at $\phi = 0$. Then we get

$$
\mathbf{h}(0) = \mathbf{0}. \quad (1.37)
$$

It follows from (1.36) that

$$
\mathbf{h}'(\phi) = -2p(\phi)\,\mathbf{H}(\phi)\, \mathbf{a} = -2p(\phi)\,\mathbf{h}(\phi). \quad (1.38)
$$

From (1.37) and the uniqueness of the solution to the initial value problem for the ordinary differential equation (1.38) we obtain

$$\mathbf{h}(\phi) = \left[\mathbf{Q} + p\left(\mathbf{R} + \mathbf{R}^T\right) + p^2\mathbf{T}\right](\phi)\,\mathbf{a} = \mathbf{0} \tag{1.39}$$

for all ϕ. This implies that $\mathbf{a}$ is the solution to (1.21) for all ϕ.

Now put

$$\eta(\phi) = \left[\mathbf{R}^T + p\,\mathbf{T}\right](\phi)\,\mathbf{a}.$$

Then, by the assumption of the theorem, we have

$$\eta(0) = \mathbf{l}.$$

It follows from (1.34) and (1.35) that

$$
\begin{aligned}
\eta'(\phi) &= \left[\mathbf{T} - \mathbf{Q} + p'\mathbf{T} - p\left(\mathbf{R} + \mathbf{R}^T\right)\right](\phi)\,\mathbf{a} \\
&= -\left[\mathbf{Q} + p\left(\mathbf{R} + \mathbf{R}^T\right) + p^2\mathbf{T}\right](\phi)\,\mathbf{a} = -\mathbf{h}(\phi),
\end{aligned}
\tag{1.40}
$$

which is equal to zero by (1.39). Hence we obtain

$$\eta(\phi) = \left[\mathbf{R}^T + p\,\mathbf{T}\right](\phi)\,\mathbf{a} = \eta(0) = \mathbf{l} \tag{1.41}$$

for all ϕ. This implies that $\mathbf{l}$ satisfies (1.25) for all ϕ.   □

Remark 1.4 The Riccati equation (1.34) can be solved explicitly as

$$p(\phi) = \tan(\phi_c - \phi),$$

where ϕ_c is a non-real constant determined by the initial condition $p(0) = p_1 = \tan\phi_c$, or

$$p(\phi) \equiv \pm\sqrt{-1}.$$

In the proof of Theorem 1.3, we have assumed that $p(\phi)$ is the solution to the Riccati equation (1.34). However, it follows that $p(\phi)$ *must be* the solution to (1.34). In fact, (1.39) and (1.41) are equivalent to the eigenvalue problem (1.29) with $\mathbf{N} = \mathbf{N}(\phi)$ given by (1.33). Then $\begin{bmatrix}\mathbf{a}\\\mathbf{l}\end{bmatrix} \in \mathbb{C}^6$ is an eigenvector of $\mathbf{N}(\phi)$ corresponding to the eigenvalue $p(\phi)$. Therefore, $p(\phi)$ is unique for each ϕ and satisfies (1.34).

Thus, we can rewrite Theorem 1.3 in terms of Stroh's eigenvalue problem (1.29) as follows:

Theorem 1.5 *Let* $\begin{bmatrix}\mathbf{a}\\\mathbf{l}\end{bmatrix}$ *be an eigenvector of* $\mathbf{N}(0)$ *associated with the eigenvalue* $p = p_1$. *Then it holds that*

$$\mathbf{N}(\phi)\begin{bmatrix}\mathbf{a}\\\mathbf{l}\end{bmatrix} = p(\phi)\begin{bmatrix}\mathbf{a}\\\mathbf{l}\end{bmatrix} \tag{1.42}$$

for all ϕ, *while the eigenvalue* $p(\phi)$ *of* $\mathbf{N}(\phi)$ *satisfies the Riccati equation (1.34) with* $p(0) = p_1$.

 Springer

1.4 Forms of Basic Solutions When Stroh's Eigenvalue Problem is Degenerate

In Section 1.2, considering solutions of the form (1.18) to the equations of equilibrium (1.17) has led us to the six-dimensional Stroh's eigenvalue problem (1.29). More precisely, the vector **a** in (1.18) becomes the first three components of an eigenvector of **N** (1.30), while the vector **l** in (1.26) becomes the last three components of that eigenvector.

On the other hand, when the six-dimensional eigenvalue problem (1.29) does not have six linearly independent eigenvectors, the notions of generalized eigenvectors and Jordan chain will be needed. Then we have to consider solutions of a different form than (1.18). Hence in this section we give the general forms of basic solutions to (1.17) according to the degeneracy of the eigenvalue problem (1.29). But here we restrict ourselves to giving a formal argument for constructing such forms of basic solutions. Later in Section 1.9, these forms will be justified by the theory of Fourier analysis (see also Remark 1.8).

Recall that the eigenvalues p_α $(1 \le \alpha \le 6)$ of the six-dimensional eigenvalue problem (1.29) are not real and they occur in complex conjugate pairs (see the paragraph after Theorem 1.2). Henceforth we take

$$\operatorname{Im} p_\alpha > 0, \qquad \alpha = 1, 2, 3,$$

where $\operatorname{Im} p$ denotes the imaginary part of p, and put

$$p_{\alpha+3} = \overline{p_\alpha} \quad (\alpha = 1, 2, 3), \tag{1.43}$$

where $\overline{p}$ denotes the complex conjugate of p.

Let **m** and **n** be orthogonal unit vectors in $\mathbb{R}^3$. We say that the eigenvalue problem (1.29) is *simple* if all the six eigenvalues p_α $(1 \le \alpha \le 6)$ are distinct. Then there exist six linearly independent eigenvectors $\begin{bmatrix} \mathbf{a}_\alpha \\ \mathbf{l}_\alpha \end{bmatrix}$ $(1 \le \alpha \le 6)$ which span $\mathbb{C}^6$. Taking account of (1.43), we may put

$$\begin{bmatrix} \mathbf{a}_{\alpha+3} \\ \mathbf{l}_{\alpha+3} \end{bmatrix} = \overline{\begin{bmatrix} \mathbf{a}_\alpha \\ \mathbf{l}_\alpha \end{bmatrix}} \quad (\alpha = 1, 2, 3). \tag{1.44}$$

In this case, by (1.18) and (1.26), the general form of the basic solution to (1.17) which describes two-dimensional[5] deformations in the half-space $\mathbf{n} \cdot \mathbf{x} \le 0$, satisfies the condition $\boldsymbol{u} = \mathbf{a} e^{-\sqrt{-1}\xi \, \mathbf{m} \cdot \mathbf{x}}$ on $\mathbf{n} \cdot \mathbf{x} = 0$ for some complex vector **a**, and decays to zero as $\mathbf{n} \cdot \mathbf{x} \longrightarrow -\infty$ is written as

$$\boldsymbol{u} = \sum_{\alpha=1}^{3} c_\alpha \, \mathbf{a}_\alpha \, e^{-\sqrt{-1}\xi(\mathbf{m} \cdot \mathbf{x} + p_\alpha \mathbf{n} \cdot \mathbf{x})}, \tag{1.45}$$

---

[5]For the meaning of "two-dimensional", see the third paragraph of Section 1.2.

🖄 Springer

and the corresponding traction on the surface $\mathbf{n} \cdot \mathbf{x} = 0$ is given by

$$t = -\sqrt{-1}\,\xi \sum_{\alpha=1}^{3} c_\alpha\, \mathbf{l}_\alpha\, e^{-\sqrt{-1}\,\xi\, \mathbf{m} \cdot \mathbf{x}}, \tag{1.46}$$

where c_α $(1 \le \alpha \le 3)$ are complex constants determined from $\sum_{\alpha=1}^{3} c_\alpha\, \mathbf{a}_\alpha = \mathbf{a}$, and $\xi > 0$.

We say that the eigenvalue problem (1.29) is *semisimple* if there are multiple eigenvalues in the six eigenvalues p_α $(1 \le \alpha \le 6)$ but there remain six linearly independent eigenvectors $\begin{bmatrix} \mathbf{a}_\alpha \\ \mathbf{l}_\alpha \end{bmatrix}$ $(1 \le \alpha \le 6)$ which span $\mathbb{C}^6$. We also use the conventions (1.43) and (1.44). In this case the general form of the basic solution and the corresponding traction on the surface $\mathbf{n} \cdot \mathbf{x} = 0$ are again given by (1.45) and (1.46), respectively.

We say that the eigenvalue problem (1.29) is *degenerate* or *non-semisimple* if there exist multiple eigenvalues in the six eigenvalues p_α and there do not exist six linearly independent eigenvectors any more.

The last case is classified into the two cases D1 and D2 according to the Jordan normal form of the 6×6 matrix $\mathbf{N}$ (1.30). (For the Jordan normal form of a matrix, see, for example, Sections 6.6, 6.9 and 7.7 of [29] and Section 1.4 of [77]).

Case D1 There exists a non-singular 6×6 matrix $\mathbf{P}$ which transforms $\mathbf{N}$ to the following Jordan normal form

$$\mathbf{P}^{-1}\mathbf{N}\mathbf{P} = \begin{bmatrix} p_1 & & & & & \\ & p_2 & 1 & & & \\ & & p_2 & & & \\ & & & \overline{p_1} & & \\ & & & & \overline{p_2} & 1 \\ & & & & & \overline{p_2} \end{bmatrix},$$

where the empty entries in the matrix of the right hand side are filled by zero. Denoting the α-th column of $\mathbf{P}$ by $\begin{bmatrix} \mathbf{a}_\alpha \\ \mathbf{l}_\alpha \end{bmatrix} \in \mathbb{C}^6$ $(1 \le \alpha \le 6)$, we see that $\begin{bmatrix} \mathbf{a}_\alpha \\ \mathbf{l}_\alpha \end{bmatrix}$ $(\alpha = 1, 2)$ are linearly independent eigenvectors associated with p_α $(\alpha = 1, 2)$, and $\begin{bmatrix} \mathbf{a}_3 \\ \mathbf{l}_3 \end{bmatrix}$ is a generalized eigenvector which satisfies

$$\mathbf{N}\begin{bmatrix} \mathbf{a}_3 \\ \mathbf{l}_3 \end{bmatrix} - p_2\begin{bmatrix} \mathbf{a}_3 \\ \mathbf{l}_3 \end{bmatrix} = \begin{bmatrix} \mathbf{a}_2 \\ \mathbf{l}_2 \end{bmatrix}. \tag{1.47}$$

Likewise we use the convention (1.44). Then all six $\begin{bmatrix} \mathbf{a}_\alpha \\ \mathbf{l}_\alpha \end{bmatrix}$ $(1 \le \alpha \le 6)$ are linearly independent and span $\mathbb{C}^6$.

Lemma 1.6 *The general form of the basic solution to* (1.17) *in Case* D1 *which describes two-dimensional deformations in the half-space* $\mathbf{n} \cdot \mathbf{x} \le 0$, *satisfies the condition*

$u = a e^{-\sqrt{-1}\xi\, \mathbf{m}\cdot\mathbf{x}}$ on $\mathbf{n}\cdot\mathbf{x} = 0$ *for some* $a \in \mathbb{C}^3$, *and decays to zero as* $\mathbf{n}\cdot\mathbf{x} \longrightarrow -\infty$ *is written as*

$$u = \sum_{\alpha=1}^{2} c_\alpha\, \mathbf{a}_\alpha\, e^{-\sqrt{-1}\xi(\mathbf{m}\cdot\mathbf{x}+p_\alpha \mathbf{n}\cdot\mathbf{x})} + c_3\left(\mathbf{a}_3 - \sqrt{-1}\,\xi(\mathbf{n}\cdot\mathbf{x})\,\mathbf{a}_2\right)e^{-\sqrt{-1}\xi(\mathbf{m}\cdot\mathbf{x}+p_2 \mathbf{n}\cdot\mathbf{x})}, \quad (1.48)$$

and the corresponding traction on the surface $\mathbf{n}\cdot\mathbf{x} = 0$ *is given by*

$$t = -\sqrt{-1}\,\xi \sum_{\alpha=1}^{3} c_\alpha\, \mathbf{l}_\alpha\, e^{-\sqrt{-1}\xi\, \mathbf{m}\cdot\mathbf{x}}, \qquad (1.49)$$

where c_α $(1 \le \alpha \le 3)$ *are complex constants determined from* $\sum_{\alpha=1}^{3} c_\alpha \mathbf{a}_\alpha = \mathbf{a}$, *and* $\xi > 0$.

Proof Looking at the first three rows of the system (1.47), from (1.30) we obtain

$$-\mathbf{T}^{-1}\mathbf{R}^T\mathbf{a}_3 + \mathbf{T}^{-1}\mathbf{l}_3 - p_2\mathbf{a}_3 = \mathbf{a}_2,$$

which gives

$$\mathbf{l}_3 = \left[\mathbf{R}^T + p_2\mathbf{T}\right]\mathbf{a}_3 + \mathbf{T}\,\mathbf{a}_2. \qquad (1.50)$$

Looking at the last three rows, we get

$$\left[-\mathbf{Q} + \mathbf{R}\mathbf{T}^{-1}\mathbf{R}^T\right]\mathbf{a}_3 - \mathbf{R}\mathbf{T}^{-1}\mathbf{l}_3 - p_2\mathbf{l}_3 = \mathbf{l}_2.$$

Then substituting (1.50) into the preceding equation, we get

$$\left[\mathbf{Q} + p_2\left(\mathbf{R} + \mathbf{R}^T\right) + p_2^2\mathbf{T}\right]\mathbf{a}_3 + \left[\mathbf{R} + p_2\mathbf{T}\right]\mathbf{a}_2 + \mathbf{l}_2 = \mathbf{0}. \qquad (1.51)$$

Noting that $\begin{bmatrix}\mathbf{a}_2\\\mathbf{l}_2\end{bmatrix}$ is an eigenvector of $\mathbf{N}$, which satisfies

$$\mathbf{l}_2 = \left[\mathbf{R}^T + p_2\mathbf{T}\right]\mathbf{a}_2,$$

we obtain

$$\left[\mathbf{Q} + p_2\left(\mathbf{R} + \mathbf{R}^T\right) + p_2^2\mathbf{T}\right]\mathbf{a}_3 + \left[\mathbf{R} + \mathbf{R}^T + 2p_2\mathbf{T}\right]\mathbf{a}_2 = \mathbf{0}. \qquad (1.52)$$

Hence (1.47) is equivalent to (1.50) and (1.52). To see that (1.48) satisfies (1.17), we note that the first two terms $\sum_{\alpha=1}^{2} c_\alpha \mathbf{a}_\alpha e^{-\sqrt{-1}\xi(\mathbf{m}\cdot\mathbf{x}+p_\alpha \mathbf{n}\cdot\mathbf{x})}$ in the right hand side of (1.48) observe (1.17), because $\mathbf{a}_\alpha$ $(\alpha = 1, 2)$ are the same as the displacement part of a Stroh eigenvector in Section 1.2. Hence it remains to prove that the last term $\left(\mathbf{a}_3 - \sqrt{-1}\,\xi(\mathbf{n}\cdot\mathbf{x})\,\mathbf{a}_2\right)e^{-\sqrt{-1}\xi(\mathbf{m}\cdot\mathbf{x}+p_2 \mathbf{n}\cdot\mathbf{x})}$ satisfies (1.17). Putting

$$\tilde{u} = (\tilde{u}_1, \tilde{u}_2, \tilde{u}_3) = \left(\mathbf{a}_3 - \sqrt{-1}\,\xi(\mathbf{n}\cdot\mathbf{x})\,\mathbf{a}_2\right)e^{-\sqrt{-1}\xi(\mathbf{m}\cdot\mathbf{x}+p_2 \mathbf{n}\cdot\mathbf{x})}, \qquad (1.53)$$

we obtain

$$\sum_{j,k,l=1}^{3} \left(C_{ijkl} \frac{\partial^2 \tilde{u}_k}{\partial x_j \partial x_l} \right)_{i\downarrow 1,2,3}$$

$$= -\xi^2 \left(\sum_{j,l=1}^{3} C_{ijkl} \, m_j m_l + p_2 \left(\sum_{j,l=1}^{3} C_{ijkl} \, m_j n_l + \sum_{j,l=1}^{3} C_{ijkl} \, n_j m_l \right) \right.$$

$$\left. + p_2^2 \sum_{j,l=1}^{3} C_{ijkl} \, n_j n_l \right)_{i\downarrow k \to 1,2,3} \left(\mathbf{a}_3 - \sqrt{-1}\,\xi\,(\mathbf{n}\cdot\mathbf{x})\,\mathbf{a}_2 \right) e^{-\sqrt{-1}\,\xi(\mathbf{m}\cdot\mathbf{x}+p_2\mathbf{n}\cdot\mathbf{x})}$$

$$- \xi^2 \left(\sum_{j,l=1}^{3} C_{ijkl} \left(n_j(m_l+p_2 n_l)+n_l(m_j+p_2 n_j) \right) \right)_{i\downarrow k \to 1,2,3} \mathbf{a}_2 \, e^{-\sqrt{-1}\,\xi(\mathbf{m}\cdot\mathbf{x}+p_2\mathbf{n}\cdot\mathbf{x})},$$

and from (1.20) and (1.21) for $\mathbf{a} = \mathbf{a}_2$,

$$= \xi^2 \left(-\left[\mathbf{Q} + p_2 \left(\mathbf{R}+\mathbf{R}^T \right) + p_2^2 \mathbf{T} \right] \mathbf{a}_3 - \left[\mathbf{R}+\mathbf{R}^T + 2p_2\mathbf{T} \right] \mathbf{a}_2 \right) e^{-\sqrt{-1}\,\xi(\mathbf{m}\cdot\mathbf{x}+p_2\mathbf{n}\cdot\mathbf{x})},$$

which is equal to zero by (1.52).

Now we will show (1.49). By (1.25) and (1.26), the first two terms $-\sqrt{-1}\,\xi$ $\sum_{\alpha=1}^{2} c_\alpha \, \mathbf{l}_\alpha \, e^{-\sqrt{-1}\,\xi\,\mathbf{m}\cdot\mathbf{x}}$ in the right hand side of (1.49) are the traction on the surface $\mathbf{n}\cdot\mathbf{x} = 0$ produced by $\sum_{\alpha=1}^{2} c_\alpha \, \mathbf{a}_\alpha \, e^{-\sqrt{-1}(\mathbf{m}\cdot\mathbf{x}+p_\alpha\mathbf{n}\cdot\mathbf{x})}$ in (1.48). Hence we will compute the traction on $\mathbf{n}\cdot\mathbf{x} = 0$ produced by (1.53). It follows that

$$\left(\sum_{j,k,l=1}^{3} C_{ijkl} \frac{\partial \tilde{u}_k}{\partial x_l} n_j \right)_{i\downarrow 1,2,3}$$

$$= -\sqrt{-1}\,\xi \left(\sum_{j,l=1}^{3} C_{ijkl} \, (m_l + p_2 n_l) \, n_j \right)_{i\downarrow k \to 1,2,3}$$

$$\times \left(\mathbf{a}_3 - \sqrt{-1}\,\xi\,(\mathbf{n}\cdot\mathbf{x})\,\mathbf{a}_2 \right) e^{-\sqrt{-1}\,\xi(\mathbf{m}\cdot\mathbf{x}+p_2\mathbf{n}\cdot\mathbf{x})}$$

$$- \sqrt{-1}\,\xi \left(\sum_{j,l=1}^{3} C_{ijkl} \, n_j n_l \right)_{i\downarrow k \to 1,2,3} \mathbf{a}_2 \, e^{-\sqrt{-1}\,\xi(\mathbf{m}\cdot\mathbf{x}+p_2\mathbf{n}\cdot\mathbf{x})}$$

$$= -\sqrt{-1}\,\xi \left[\mathbf{R}^T + p_2\mathbf{T} \right] \left(\mathbf{a}_3 - \sqrt{-1}\,\xi\,(\mathbf{n}\cdot\mathbf{x})\,\mathbf{a}_2 \right) e^{-\sqrt{-1}\,\xi(\mathbf{m}\cdot\mathbf{x}+p_2\mathbf{n}\cdot\mathbf{x})}$$

$$- \sqrt{-1}\,\xi\,\mathbf{T}\,\mathbf{a}_2 \, e^{-\sqrt{-1}\,\xi(\mathbf{m}\cdot\mathbf{x}+p_2\mathbf{n}\cdot\mathbf{x})}.$$

This is written on the surface $\mathbf{n}\cdot\mathbf{x} = 0$ as

$$-\sqrt{-1}\,\xi \left(\left[\mathbf{R}^T + p_2\mathbf{T} \right] \mathbf{a}_3 + \mathbf{T}\,\mathbf{a}_2 \right) e^{-\sqrt{-1}\,\xi\,\mathbf{m}\cdot\mathbf{x}},$$

which is equal to

$$-\sqrt{-1}\,\xi\,\mathbf{l}_3 \, e^{-\sqrt{-1}\,\xi\,\mathbf{m}\cdot\mathbf{x}}$$

by (1.50). $\qquad\square$

Now we proceed to the Case D2.

Case D2 There exists a non-singular 6×6 matrix $\mathbf{P}$ which transforms $\mathbf{N}$ to the following Jordan normal form

$$\mathbf{P}^{-1}\mathbf{N}\mathbf{P} = \begin{bmatrix} p_1 & 1 & & & & \\ & p_1 & 1 & & & \\ & & p_1 & & & \\ & & & \overline{p_1} & 1 & \\ & & & & \overline{p_1} & 1 \\ & & & & & \overline{p_1} \end{bmatrix}.$$

Denoting the α-th column of $\mathbf{P}$ by $\begin{bmatrix} \mathbf{a}_\alpha \\ \mathbf{l}_\alpha \end{bmatrix} \in \mathbb{C}^6$ ($1 \leq \alpha \leq 6$), we see that $\begin{bmatrix} \mathbf{a}_1 \\ \mathbf{l}_1 \end{bmatrix}$ is an eigenvector associated with the triple eigenvalue p_1, and $\begin{bmatrix} \mathbf{a}_\alpha \\ \mathbf{l}_\alpha \end{bmatrix}$ ($\alpha = 2, 3$) are generalized eigenvectors which satisfy

$$\mathbf{N}\begin{bmatrix} \mathbf{a}_2 \\ \mathbf{l}_2 \end{bmatrix} - p_1 \begin{bmatrix} \mathbf{a}_2 \\ \mathbf{l}_2 \end{bmatrix} = \begin{bmatrix} \mathbf{a}_1 \\ \mathbf{l}_1 \end{bmatrix}, \qquad \mathbf{N}\begin{bmatrix} \mathbf{a}_3 \\ \mathbf{l}_3 \end{bmatrix} - p_1 \begin{bmatrix} \mathbf{a}_3 \\ \mathbf{l}_3 \end{bmatrix} = \begin{bmatrix} \mathbf{a}_2 \\ \mathbf{l}_2 \end{bmatrix}. \tag{1.54}$$

As before, we use the convention (1.44). Then all six $\begin{bmatrix} \mathbf{a}_\alpha \\ \mathbf{l}_\alpha \end{bmatrix}$ ($1 \leq \alpha \leq 6$) are linearly independent and span $\mathbb{C}^6$.[6]

Lemma 1.7 *The general form of the basic solution to* (1.17) *in Case* D2 *which describes two-dimensional deformations in the half-space* $\mathbf{n} \cdot \mathbf{x} \leq 0$, *satisfies the condition* $\boldsymbol{u} = \mathbf{a}\,e^{-\sqrt{-1}\,\xi\,\mathbf{m}\cdot\mathbf{x}}$ *on* $\mathbf{n} \cdot \mathbf{x} = 0$ *for some* $\mathbf{a} \in \mathbb{C}^3$, *and decays to zero as* $\mathbf{n} \cdot \mathbf{x} \longrightarrow -\infty$ *is written as*

$$\boldsymbol{u} = \left(c_1\,\mathbf{a}_1 + c_2 \left(\mathbf{a}_2 - \sqrt{-1}\,\xi\,(\mathbf{n} \cdot \mathbf{x})\,\mathbf{a}_1 \right) \right.$$
$$\left. + c_3 \left(\mathbf{a}_3 - \sqrt{-1}\,\xi\,(\mathbf{n} \cdot \mathbf{x})\,\mathbf{a}_2 - \frac{1}{2}\xi^2\,(\mathbf{n} \cdot \mathbf{x})^2\,\mathbf{a}_1 \right) \right) e^{-\sqrt{-1}\,\xi(\mathbf{m}\cdot\mathbf{x}+p_1\mathbf{n}\cdot\mathbf{x})}, \tag{1.55}$$

---

[6]We say that an eigenvector $\mathbf{v}_1$ and generalized eigenvectors $\mathbf{v}_2, \cdots, \mathbf{v}_k$ associated with the multiple eigenvalue λ of the $n \times n$ matrix $\mathbf{M}$ form a Jordan chain of length k ($k \leq n$) if

$$\mathbf{M}\mathbf{v}_1 = \lambda\,\mathbf{v}_1, \qquad \mathbf{M}\mathbf{v}_i - \lambda\mathbf{v}_i = \mathbf{v}_{i-1}, \qquad i = 2, 3, \cdots, k.$$

The corresponding Jordan block is a $k \times k$ matrix with repeated eigenvalue λ on the diagonal, 1 on the first super diagonal and zero elsewhere. Note that although the eigenvalue problem (1.29) is six-dimensional, the length of the Jordan chains is at most three, because the eigenvalues are not real and occur in complex conjugate pairs.

Thus $\begin{bmatrix} \mathbf{a}_\alpha \\ \mathbf{l}_\alpha \end{bmatrix}$ ($\alpha = 2, 3$) in Case D1 form a Jordan chain of length 2, while $\begin{bmatrix} \mathbf{a}_\alpha \\ \mathbf{l}_\alpha \end{bmatrix}$ ($\alpha = 1, 2, 3$) in Case D2 form a Jordan chain of length 3.

and the corresponding traction on the surface $\mathbf{n} \cdot \mathbf{x} = 0$ *is given by*

$$t = -\sqrt{-1}\,\xi \sum_{\alpha=1}^{3} c_\alpha \, \mathbf{l}_\alpha \, e^{-\sqrt{-1}\,\xi\,\mathbf{m}\cdot\mathbf{x}}, \tag{1.56}$$

where c_α $(1 \le \alpha \le 3)$ *are complex constants determined from* $\sum_{\alpha=1}^{3} c_\alpha \, \mathbf{a}_\alpha = \mathbf{a}$, *and* $\xi > 0$.

This lemma can be proved in a similar way as that for Lemma 1.6. Namely, we write down from the first three rows and the last three rows of the systems (1.54) the four systems in $\mathbb{C}^3$ equivalent to (1.54). On the other hand, we substitute (1.55) into the left hand side of (1.17) and (1.24). Then we compare the resulting formulas. The details are left as Exercise 1-4.

Remark 1.8 Although the forms of basic solutions (displacements) are different according to the degeneracy of the eigenvalue problem (1.29) (see (1.45), (1.48) and (1.55)), they have the same form on the surface $\mathbf{n} \cdot \mathbf{x} = 0$, namely

$$\boldsymbol{u} = \sum_{\alpha=1}^{3} c_\alpha \, \mathbf{a}_\alpha \, e^{-\sqrt{-1}\,\xi\,\mathbf{m}\cdot\mathbf{x}}. \tag{1.57}$$

The corresponding tractions on $\mathbf{n} \cdot \mathbf{x} = 0$ also assume the same form

$$t = -\sqrt{-1}\,\xi \sum_{\alpha=1}^{3} c_\alpha \, \mathbf{l}_\alpha \, e^{-\sqrt{-1}\,\xi\,\mathbf{m}\cdot\mathbf{x}}$$

(see (1.46), (1.49) and (1.56)).

Looking back on the arguments in this section, we observe that formulas (1.45), (1.48) and (1.55) give, for two-dimensional deformations in the half-space $\mathbf{n} \cdot \mathbf{x} \le 0$, the forms of the solution for the elementary boundary condition (1.57) on $\mathbf{n} \cdot \mathbf{x} = 0$ according to the degeneracy of the eigenvalue problem (1.29). In Section 1.9 we shall see that considering boundary condition (1.57) suffices for the boundary-value problem in the half-space $\mathbf{n} \cdot \mathbf{x} \le 0$ and the above forms of the basic solution will be justified by the theory of Fourier analysis.

From now on, we call the vectors $\mathbf{a}_\alpha \in \mathbb{C}^3$ ($\alpha = 1, 2, 3$) the displacement parts and $\mathbf{l}_\alpha \in \mathbb{C}^3$ ($\alpha = 1, 2, 3$) the traction parts of $\begin{bmatrix} \mathbf{a}_\alpha \\ \mathbf{l}_\alpha \end{bmatrix} \in \mathbb{C}^6$, not only when $\begin{bmatrix} \mathbf{a}_\alpha \\ \mathbf{l}_\alpha \end{bmatrix}$ is an eigenvector of $\mathbf{N}$ corresponding to an eigenvalue with positive imaginary part, but also when $\begin{bmatrix} \mathbf{a}_\alpha \\ \mathbf{l}_\alpha \end{bmatrix}$ is a generalized eigenvector of $\mathbf{N}$ corresponding to an eigenvalue with positive imaginary part.

1.5 Rotational Dependence When Stroh's Eigenvalue Problem is Degenerate

In Section 1.3 we have proved that eigenvectors of $\mathbf{N}$ are invariant under the rotations of $\mathbf{m}$ and $\mathbf{n}$ around the vector product $\mathbf{m} \times \mathbf{n}$ (Theorem 1.5). In this section we consider dependence of generalized eigenvectors of $\mathbf{N}$ on the rotations of $\mathbf{m}$ and $\mathbf{n}$ when Stroh's eigenvalue problem (1.29) is degenerate. As we shall see, generalized

 Springer

eigenvectors do not enjoy such elegant invariance as in Section 1.3, but there is a rule that describes their dependence on the rotations of $\mathbf{m}$ and $\mathbf{n}$.

As in Section 1.3, we consider here rotations given by (1.31) with rotation angle ϕ and use notations in (1.32) and (1.33).

Corresponding to the Case D1 in Section 1.4, we have

Theorem 1.9 *Let* $\begin{bmatrix} \mathbf{a}_\alpha \\ \mathbf{l}_\alpha \end{bmatrix}$ *($\alpha = 1, 2$) be linearly independent eigenvectors of* $\mathbf{N}(0)$ *associated with the eigenvalues* p_α *($\alpha = 1, 2$), and let* $\begin{bmatrix} \mathbf{a}_3 \\ \mathbf{l}_3 \end{bmatrix}$ *be a generalized eigenvector satisfying*

$$\mathbf{N}(0)\begin{bmatrix} \mathbf{a}_3 \\ \mathbf{l}_3 \end{bmatrix} - p_2 \begin{bmatrix} \mathbf{a}_3 \\ \mathbf{l}_3 \end{bmatrix} = \begin{bmatrix} \mathbf{a}_2 \\ \mathbf{l}_2 \end{bmatrix}. \tag{1.58}$$

Then it follows that

$$\mathbf{N}(\phi)\begin{bmatrix} \mathbf{a}_\alpha \\ \mathbf{l}_\alpha \end{bmatrix} = p_\alpha(\phi)\begin{bmatrix} \mathbf{a}_\alpha \\ \mathbf{l}_\alpha \end{bmatrix}, \qquad \alpha = 1, 2 \tag{1.59}$$

for all ϕ, *and by putting*

$$\begin{bmatrix} \mathbf{a}_3(\phi) \\ \mathbf{l}_3(\phi) \end{bmatrix} = \exp\left(\int_0^\phi 2 p_2(\psi)\, d\psi\right)\begin{bmatrix} \mathbf{a}_3 \\ \mathbf{l}_3 \end{bmatrix}, \tag{1.60}$$

it follows that

$$\mathbf{N}(\phi)\begin{bmatrix} \mathbf{a}_3(\phi) \\ \mathbf{l}_3(\phi) \end{bmatrix} - p_2(\phi)\begin{bmatrix} \mathbf{a}_3(\phi) \\ \mathbf{l}_3(\phi) \end{bmatrix} = \begin{bmatrix} \mathbf{a}_2 \\ \mathbf{l}_2 \end{bmatrix} \tag{1.61}$$

for all ϕ, *where* $p_\alpha(\phi)$ *($\alpha = 1, 2$) are the eigenvalues of* $\mathbf{N}(\phi)$ *and satisfy the Riccati equation* (1.34) *with* $p_\alpha(0) = p_\alpha$ *($\alpha = 1, 2$).*

Proof Equations (1.59) are immediate consequences of Theorem 1.5. Now we will prove (1.61). Recalling that the eigenrelation (1.47) is equivalent to (1.50) and (1.51), we only have to show that

$$\left[\mathbf{Q} + p_2\left(\mathbf{R} + \mathbf{R}^T\right) + p_2^2\mathbf{T}\right](\phi)\,\mathbf{a}_3(\phi) + \left[\mathbf{R} + p_2\mathbf{T}\right](\phi)\,\mathbf{a}_2 + \mathbf{l}_2 = \mathbf{0} \tag{1.62}$$

and

$$\mathbf{l}_3(\phi) - \left[\mathbf{R}^T + p_2\mathbf{T}\right](\phi)\,\mathbf{a}_3(\phi) - \mathbf{T}(\phi)\,\mathbf{a}_2 = \mathbf{0} \tag{1.63}$$

for all ϕ. Put the left hand sides of (1.62) and (1.63) by $\mathbf{g}(\phi)$ and $\zeta(\phi)$, respectively. Then by (1.58), we get

$$\mathbf{g}(0) = \mathbf{0} \quad \text{and} \quad \zeta(0) = \mathbf{0}. \tag{1.64}$$

Since it follows from (1.60) that

$$\mathbf{a}_3'(\phi) = 2 p_2(\phi)\,\mathbf{a}_3(\phi) \quad \text{and} \quad \mathbf{l}_3'(\phi) = 2 p_2(\phi)\,\mathbf{l}_3(\phi), \tag{1.65}$$

from (1.36), (1.40), and (1.39) for $\mathbf{a} = \mathbf{a}_2$ we see that

$$\mathbf{g}'(\phi) = \mathbf{0},$$

which, when combined with (1.64), implies that

$$\mathbf{g}(\phi) = \mathbf{0}$$

for all ϕ.

In the same way, from (1.35), (1.40) and (1.65) we obtain

$$\zeta'(\phi) = 2\, p_2(\phi)\, \mathbf{l}_3(\phi) + \left[\mathbf{Q} + p_2(\mathbf{R} + \mathbf{R}^T) + p_2^2\mathbf{T}\right](\phi)\, \mathbf{a}_3(\phi)$$

$$- 2\, p_2(\phi)\left[\mathbf{R}^T + p_2\mathbf{T}\right](\phi)\, \mathbf{a}_3(\phi) + \left[\mathbf{R} + \mathbf{R}^T\right](\phi)\, \mathbf{a}_2$$

and using (1.62) in the second term of the right hand side, we get

$$\zeta'(\phi) = 2\, p_2(\phi)\left(\mathbf{l}_3(\phi) - \left[\mathbf{R}^T + p_2\mathbf{T}\right](\phi)\, \mathbf{a}_3(\phi)\right) + \left[\mathbf{R}^T - p_2\mathbf{T}\right](\phi)\, \mathbf{a}_2 - \mathbf{l}_2,$$

and then, from (1.41) with $\mathbf{a} = \mathbf{a}_2, \mathbf{l} = \mathbf{l}_2$ we have

$$\zeta'(\phi) = 2\, p_2(\phi)\, \zeta(\phi),$$

which, when combined with (1.64), implies that

$$\zeta(\phi) = \mathbf{0}$$

for all ϕ. □

Corresponding to the Case D2, we have

Theorem 1.10 *Let* $\begin{bmatrix} \mathbf{a}_1 \\ \mathbf{l}_1 \end{bmatrix}$ *be an eigenvector of* $\mathbf{N}(0)$ *associated with the triple eigenvalue* p_1, *and let* $\begin{bmatrix} \mathbf{a}_\alpha \\ \mathbf{l}_\alpha \end{bmatrix}$ *($\alpha = 2, 3$) be generalized eigenvectors satisfying*

$$\mathbf{N}(0)\begin{bmatrix} \mathbf{a}_2 \\ \mathbf{l}_2 \end{bmatrix} - p_1\begin{bmatrix} \mathbf{a}_2 \\ \mathbf{l}_2 \end{bmatrix} = \begin{bmatrix} \mathbf{a}_1 \\ \mathbf{l}_1 \end{bmatrix}, \qquad \mathbf{N}(0)\begin{bmatrix} \mathbf{a}_3 \\ \mathbf{l}_3 \end{bmatrix} - p_1\begin{bmatrix} \mathbf{a}_3 \\ \mathbf{l}_3 \end{bmatrix} = \begin{bmatrix} \mathbf{a}_2 \\ \mathbf{l}_2 \end{bmatrix}.$$

Then it follows that

$$\mathbf{N}(\phi)\begin{bmatrix} \mathbf{a}_1 \\ \mathbf{l}_1 \end{bmatrix} = p_1(\phi)\begin{bmatrix} \mathbf{a}_1 \\ \mathbf{l}_1 \end{bmatrix}$$

for all ϕ, and by putting

$$\begin{bmatrix} \mathbf{a}_2(\phi) \\ \mathbf{l}_2(\phi) \end{bmatrix} = \exp\left(\int_0^\phi 2\, p_1(\psi)\, d\psi\right)$$

$$\times \left(\begin{bmatrix} \mathbf{a}_2 \\ \mathbf{l}_2 \end{bmatrix} - \int_0^\phi \exp\left(-\int_0^\psi 2\, p_1(\theta)\, d\theta\right) d\psi \begin{bmatrix} \mathbf{a}_1 \\ \mathbf{l}_1 \end{bmatrix}\right)$$

and

$$\begin{bmatrix} \mathbf{a}_3(\phi) \\ \mathbf{l}_3(\phi) \end{bmatrix} = \exp\left(\int_0^\phi 4\, p_1(\psi)\, d\psi\right)\begin{bmatrix} \mathbf{a}_3 \\ \mathbf{l}_3 \end{bmatrix},$$

it follows that

$$\mathbf{N}(\phi)\begin{bmatrix} \mathbf{a}_2(\phi) \\ \mathbf{l}_2(\phi) \end{bmatrix} - p_1(\phi)\begin{bmatrix} \mathbf{a}_2(\phi) \\ \mathbf{l}_2(\phi) \end{bmatrix} = \begin{bmatrix} \mathbf{a}_1 \\ \mathbf{l}_1 \end{bmatrix}$$

 Springer

and

$$\mathbf{N}(\phi)\begin{bmatrix}\mathbf{a}_3(\phi)\\\mathbf{l}_3(\phi)\end{bmatrix} - p_1(\phi)\begin{bmatrix}\mathbf{a}_3(\phi)\\\mathbf{l}_3(\phi)\end{bmatrix} = \begin{bmatrix}\mathbf{a}_2(\phi)\\\mathbf{l}_2(\phi)\end{bmatrix}$$

for all ϕ, where $p_1(\phi)$ is the eigenvalue of $\mathbf{N}(\phi)$ that satisfies the Riccati equation (1.34) with $p_1(0) = p_1$.

The proof can be written down in an almost parallel way to that of Theorem 1.9 but will be a little complicated. We suggest an alternative proof of this theorem in Exercise 1-7.

Finally, from Theorems 1.5, 1.9 and 1.10 we observe

Corollary 1.11 *The property that Stroh's eigenvalue problem* (1.29) *is simple, semi-simple, or degenerate remains invariant under the rotations of $\mathbf{m}$ and $\mathbf{n}$ around the vector product $\mathbf{m} \times \mathbf{n}$. Moreover, in the case where the eigenvalue problem* (1.29) *is degenerate, the structure of the Jordan chains is also invariant under those rotations.*

Therefore, it is the elasticity tensor $\mathbf{C} = (C_{ijkl})_{1 \le i,j,k,l \le 3}$ and the unit normal $\mathbf{m} \times \mathbf{n}$ to the reference plane that determine whether Stroh's eigenvalue problem (1.29) is simple, semisimple, or degenerate and, when Stroh's eigenvalue problem is degenerate, how long the Jordan chains are.

1.6 Angular Average of Stroh's Eigenvalue Problem: Integral Formalism

We have seen in Sections 1.3 and 1.5 that rotational dependency of eigenvectors and generalized eigenvectors of $\mathbf{N}$ is expressed in different forms according to the degeneracy of Stroh's eigenvalue problem and becomes more complicated as the degeneracy increases. In this section, as a completion of the study on rotational dependency of Stroh's eigenvalue problem, we take the averages of the formulas in Theorems 1.5, 1.9 and 1.10 with respect to ϕ over $[-\pi, \pi]$. Then we arrive at "the angle-averaged Stroh's eigenvalue problem", which is valid regardless of whether Stroh's eigenvalue problem is simple, semisimple or degenerate, and whose form is much simpler than the formulas in Theorems 1.9 and 1.10. Thus we obtain a fundamental theorem in the Barnett-Lothe integral formalism of Stroh's eigenvalue problem. By this theorem, linear independence of the displacement parts of Stroh eigenvectors or generalized eigenvectors will be proved, and hence we can define the surface impedance tensor in the next section.

To begin with, letting $\mathbf{N}(\phi)$ be given by (1.33), we take the angular average of its eigenvalue $p(\phi)$ over $[-\pi, \pi]$.

Lemma 1.12 *For each eigenvalue $p(\phi)$ of $\mathbf{N}(\phi)$ with $\mathrm{Im}\, p(0) > 0$,[7]*

$$\frac{1}{2\pi}\int_{-\pi}^{\pi} p(\phi)\, d\phi = \sqrt{-1}. \tag{1.66}$$

---

[7]Since $p(\phi)$ is non-real and continuous in ϕ, $\mathrm{Im}\, p(0) > 0$ is equivalent to $\mathrm{Im}\, p(\phi) > 0$ for any ϕ.

Proof According to Remark 1.4, we have $p(\phi) = \tan(\phi_c - \phi)$ with $\mathrm{Im}(\tan \phi_c) > 0$, or $p(\phi) \equiv \sqrt{-1}$. When $p(\phi) \equiv \sqrt{-1}$, we immediately get (1.66). Now let $p(\phi) = \tan(\phi_c - \phi)$ and write $\phi_c = \phi_r + \sqrt{-1}\,\phi_i$ ($\phi_r, \phi_i \in \mathbb{R}$). From $\mathrm{Im}(\tan \phi_c) > 0$ we have $\phi_i > 0$.[8] Then it follows that[9]

$$\frac{1}{2\pi} \int_{-\pi}^{\pi} p(\phi)\, d\phi = \frac{1}{2\pi} \Big[\log \cos(\phi_c - \phi) \Big]_{\phi=-\pi}^{\phi=\pi}$$

$$= \frac{1}{2\pi} \Big[\log \Big(\cos(\phi - \phi_r) \cosh \phi_i + \sqrt{-1} \sin(\phi - \phi_r) \sinh \phi_i \Big) \Big]_{\phi=-\pi}^{\phi=\pi}.$$

The argument of the complex valued $\cos(\phi - \phi_r) \cosh \phi_i + \sqrt{-1} \sin(\phi - \phi_r) \sinh \phi_i$ increases by 2π when ϕ changes from $-\pi$ to π, because $\phi_i > 0$ and $\sinh \phi_i > 0$. This proves (1.66). $\square$

Definition 1.13 We define the 6×6 real matrix $\mathbf{S}$ to be the angular average of the 6×6 matrix $\mathbf{N}(\phi)$ over $[-\pi, \pi]$:

$$\mathbf{S} = \begin{bmatrix} \mathbf{S}_1 & \mathbf{S}_2 \\ \mathbf{S}_3 & \mathbf{S}_1^T \end{bmatrix} = \frac{1}{2\pi} \int_{-\pi}^{\pi} \mathbf{N}(\phi)\, d\phi, \tag{1.67}$$

where $\mathbf{S}_1, \mathbf{S}_2$ and $\mathbf{S}_3$ are 3×3 real matrices defined by

$$\mathbf{S}_1 = \frac{1}{2\pi} \int_{-\pi}^{\pi} -\mathbf{T}(\phi)^{-1} \mathbf{R}(\phi)^T\, d\phi, \qquad \mathbf{S}_2 = \frac{1}{2\pi} \int_{-\pi}^{\pi} \mathbf{T}(\phi)^{-1}\, d\phi,$$

$$\mathbf{S}_3 = \frac{1}{2\pi} \int_{-\pi}^{\pi} \Big(-\mathbf{Q}(\phi) + \mathbf{R}(\phi)\mathbf{T}(\phi)^{-1}\mathbf{R}(\phi)^T \Big)\, d\phi. \tag{1.68}$$

Lemma 1.14 *The matrices $\mathbf{S}_2$ and $\mathbf{S}_3$ are symmetric. Furthermore, $\mathbf{S}_2$ is positive definite.*[10]

Proof The matrices $\mathbf{S}_2$ and $\mathbf{S}_3$ are symmetric, because the integrands $\mathbf{T}(\phi)^{-1}$ and $-\mathbf{Q}(\phi) + \mathbf{R}(\phi)\mathbf{T}(\phi)^{-1}\mathbf{R}(\phi)^T$ are symmetric for all ϕ. By (1) of Lemma 1.1, the matrix $\mathbf{T}(\phi)$ is positive definite for all ϕ, and so is the matrix $\mathbf{T}(\phi)^{-1}$ for all ϕ. Hence its angular average, $\mathbf{S}_2$, is also positive definite. $\square$

Now we take the angular average of Stroh's eigenvalue problem.

Theorem 1.15 *Let $\begin{bmatrix} \mathbf{a}_\alpha \\ \mathbf{l}_\alpha \end{bmatrix}$ be an eigenvector or a generalized eigenvector of $\mathbf{N}(0)$ corresponding to the eigenvalues p_α ($\alpha = 1, 2, 3$) with $\mathrm{Im}\, p_\alpha > 0$. Then it follows that*

$$\mathbf{S} \begin{bmatrix} \mathbf{a}_\alpha \\ \mathbf{l}_\alpha \end{bmatrix} = \sqrt{-1} \begin{bmatrix} \mathbf{a}_\alpha \\ \mathbf{l}_\alpha \end{bmatrix}. \tag{1.69}$$

---

[8] The formula $\tan(a + \sqrt{-1}\,b) = \frac{1}{2} \frac{\sin 2a + \sqrt{-1} \sinh 2b}{\cos^2 a + \sinh^2 b}$ ($a, b \in \mathbb{R}$) has been used.

[9] We use the formula $\cos(a + \sqrt{-1}\,b) = \cos a \cosh b - \sqrt{-1} \sin a \sinh b$ ($a, b \in \mathbb{R}$).

[10] $\mathbf{S}_3$ is negative definite, which we refer to the literature for a proof in Subsection 3.6.2.

Proof When $\begin{bmatrix} \mathbf{a}_\alpha \\ \mathbf{l}_\alpha \end{bmatrix}$ is an eigenvector of $\mathbf{N}(0)$, we take the angular averages of both sides of (1.42) and use (1.66), which immediately leads to (1.69). Note that this consideration applies not only to eigenvectors $\begin{bmatrix} \mathbf{a}_\alpha \\ \mathbf{l}_\alpha \end{bmatrix}$ when Stroh's eigenvalue problem is simple or semisimple, but also to eigenvectors of degenerate Stroh's eigenvalue problem.

Now we consider the cases where Stroh's eigenvalue problem (1.29) is degenerate, and consider Case D1 first. Let $\begin{bmatrix} \mathbf{a}_3 \\ \mathbf{l}_3 \end{bmatrix}$ be a generalized eigenvector of $\mathbf{N}(0)$ in Theorem 1.9 corresponding to the multiple eigenvalue p_2 with Im $p_2 > 0$. Then from (1.60) and (1.61) it follows that

$$\mathbf{N}(\phi)\begin{bmatrix} \mathbf{a}_3 \\ \mathbf{l}_3 \end{bmatrix} - p_2(\phi)\begin{bmatrix} \mathbf{a}_3 \\ \mathbf{l}_3 \end{bmatrix} = \exp\left(\int_0^\phi -2\,p_2(\psi)\,d\psi\right)\begin{bmatrix} \mathbf{a}_2 \\ \mathbf{l}_2 \end{bmatrix}, \qquad (1.70)$$

where $p_2(\phi)$ is the solution to the Riccati equation (1.34) with $p_2(0) = p_2$. In order to take the angular average of both sides of this equality, by Remark 1.4, two cases should be considered, i.e., the case where $p_2(\phi) = \tan(\phi_c - \phi)$ with $p_2 = \tan\phi_c$ and the case where $p_2(\phi) \equiv \sqrt{-1}$. In the case where $p_2(\phi) \equiv \sqrt{-1}$, the right hand side of (1.70) becomes

$$\exp\left(-2\sqrt{-1}\,\phi\right)\begin{bmatrix} \mathbf{a}_2 \\ \mathbf{l}_2 \end{bmatrix} = (\cos 2\phi - \sqrt{-1}\sin 2\phi)\begin{bmatrix} \mathbf{a}_2 \\ \mathbf{l}_2 \end{bmatrix}, \qquad (1.71)$$

whose angular average over $[-\pi, \pi]$ is equal to zero.

In the case where $p(\phi) = \tan(\phi_c - \phi)$, the angular average of the right hand side of (1.70) over $[-\pi, \pi]$ becomes

$$\frac{1}{2\pi}\int_{-\pi}^{\pi}\exp\left(-2\left[\log\cos(\phi_c - \psi)\right]_{\psi=0}^{\psi=\phi}\right)d\phi\begin{bmatrix} \mathbf{a}_2 \\ \mathbf{l}_2 \end{bmatrix}$$

$$= \frac{1}{2\pi}\int_{-\pi}^{\pi}\exp\left(2\log\cos\phi_c - 2\log\cos(\phi_c - \phi)\right)d\phi\begin{bmatrix} \mathbf{a}_2 \\ \mathbf{l}_2 \end{bmatrix}$$

$$= \frac{1}{2\pi}\int_{-\pi}^{\pi}\frac{\cos^2\phi_c}{\cos^2(\phi_c - \phi)}d\phi\begin{bmatrix} \mathbf{a}_2 \\ \mathbf{l}_2 \end{bmatrix} = \frac{-1}{2\pi}\cos^2\phi_c\left[\tan(\phi_c - \phi)\right]_{\phi=-\pi}^{\phi=\pi}\begin{bmatrix} \mathbf{a}_2 \\ \mathbf{l}_2 \end{bmatrix}, (1.72)$$

which is equal to zero, because Im $\phi_c > 0$ and the function $\tan(\phi_c - \phi)$ is 2π-periodic in $\phi \in \mathbb{R}$.

In both cases, the angular average of the left hand side of (1.70) over $[-\pi, \pi]$ becomes, by (1.66),

$$\mathbf{S}\begin{bmatrix} \mathbf{a}_3 \\ \mathbf{l}_3 \end{bmatrix} - \sqrt{-1}\begin{bmatrix} \mathbf{a}_\alpha \\ \mathbf{l}_\alpha \end{bmatrix}.$$

This proves (1.69).

By a similar method, the relation (1.69) can be proved for Case D2 on the basis of Theorem 1.10. However, later in Subsection 2.2.2, we shall give a systematic proof of (1.69), which is valid whatever length the Jordan chains in the eigenvalue problem (1.29) have.   □

From this theorem we conclude that the angle-averaged Stroh's eigenvalue problem has the same form (1.69) regardless of whether Stroh's eigenvalue problem (1.29) is simple, semisimple, or degenerate.

1.7 Surface Impedance Tensor

Let $\mathbf{m}$ and $\mathbf{n}$ be two orthogonal unit vectors in $\mathbb{R}^3$ and let $\begin{bmatrix} \mathbf{a}_\alpha \\ \mathbf{l}_\alpha \end{bmatrix}$ ($\alpha = 1, 2, 3$) be linearly independent eigenvector(s) or generalized eigenvector(s) of Stroh's eigenvalue problem (1.29) associated with the eigenvalues p_α ($\alpha = 1, 2, 3$, Im $p_\alpha > 0$). In this section, on the basis of Theorem 1.15, we first prove that the displacement parts $\mathbf{a}_\alpha$ ($\alpha = 1, 2, 3$) of $\begin{bmatrix} \mathbf{a}_\alpha \\ \mathbf{l}_\alpha \end{bmatrix}$ above are linearly independent.[11] Then we can define the surface impedance tensor which, for an elastic half-space, maps the displacements given at the boundary surface to the tractions needed to sustain them. This tensor, which plays a fundamental role in applications of the Stroh formalism, has been used in many studies on forward and inverse problems. In the next chapter we will give several applications of this tensor.

Now from Theorem 1.15, the first three rows of the system (1.69) is written, by using notations in (1.68), as

$$\mathbf{S}_1 \mathbf{a}_\alpha + \mathbf{S}_2 \mathbf{l}_\alpha = \sqrt{-1}\, \mathbf{a}_\alpha, \qquad \alpha = 1, 2, 3.$$

Then we get

$$\mathbf{S}_2 \mathbf{l}_\alpha = \left(\sqrt{-1}\, \mathbf{I} - \mathbf{S}_1 \right) \mathbf{a}_\alpha, \qquad \alpha = 1, 2, 3,$$

where $\mathbf{I}$ is the 3×3 identity matrix. Since Lemma 1.14 implies that $\mathbf{S}_2$ is invertible, multiplying both sides by $\mathbf{S}_2^{-1}$, we obtain

$$\mathbf{l}_\alpha = \left(\sqrt{-1}\, \mathbf{S}_2^{-1} - \mathbf{S}_2^{-1} \mathbf{S}_1 \right) \mathbf{a}_\alpha, \qquad \alpha = 1, 2, 3. \tag{1.73}$$

Suppose that $\mathbf{a}_\alpha$ ($\alpha = 1, 2, 3$) are linearly dependent. Then there exists a set of complex numbers $(c_1, c_2, c_3) \neq (0, 0, 0)$ such that

$$\sum_{\alpha=1}^{3} c_\alpha \mathbf{a}_\alpha = \mathbf{0}.$$

Then from (1.73) it follows that

$$\sum_{\alpha=1}^{3} c_\alpha \mathbf{l}_\alpha = \mathbf{0},$$

and therefore

$$\sum_{\alpha=1}^{3} c_\alpha \begin{bmatrix} \mathbf{a}_\alpha \\ \mathbf{l}_\alpha \end{bmatrix} = \mathbf{0}.$$

---

[11] Generally, if $\mathbf{a}_\alpha$ ($\alpha = 1, 2, 3$) are linearly independent in $\mathbb{C}^3$, then $\begin{bmatrix} \mathbf{a}_\alpha \\ \mathbf{l}_\alpha \end{bmatrix}$ ($\alpha = 1, 2, 3$) are linearly independent in $\mathbb{C}^6$. However, its converse does not always hold.

This contradicts with the assumption that $\begin{bmatrix} \mathbf{a}_\alpha \\ \mathbf{l}_\alpha \end{bmatrix}$ $(\alpha = 1, 2, 3)$ are linearly independent. Hence we obtain

Theorem 1.16 *Let* $\begin{bmatrix} \mathbf{a}_\alpha \\ \mathbf{l}_\alpha \end{bmatrix}$ $(\alpha = 1, 2, 3)$ *be linearly independent eigenvector(s) or generalized eigenvector(s) of Stroh's eigenvalue problem (1.29) associated with the eigenvalues* p_α $(\alpha = 1, 2, 3,\ \mathrm{Im}\ p_\alpha > 0)$. *Then their displacement parts* $\mathbf{a}_\alpha$ $(\alpha = 1, 2, 3)$ *are linearly independent.*

Thus, we are now in a position to define the surface impedance tensor.

Definition 1.17 The surface impedance tensor $\mathbf{Z}$ is the 3×3 matrix given by

$$\mathbf{Z} = -\sqrt{-1}\,[\mathbf{l}_1, \mathbf{l}_2, \mathbf{l}_3][\mathbf{a}_1, \mathbf{a}_2, \mathbf{a}_3]^{-1}, \tag{1.74}$$

where $[\mathbf{l}_1, \mathbf{l}_2, \mathbf{l}_3]$ and $[\mathbf{a}_1, \mathbf{a}_2, \mathbf{a}_3]$ denote 3×3 matrices which consist of the column vectors $\mathbf{l}_\alpha$ and $\mathbf{a}_\alpha$, respectively.

From this definition it follows that

$$\mathbf{l}_\alpha = \sqrt{-1}\,\mathbf{Z}\,\mathbf{a}_\alpha, \qquad \alpha = 1, 2, 3. \tag{1.75}$$

Therefore, taking account of Remark 1.8, we see that $\mathbf{Z}$ expresses a linear relationship between (i) the displacements given at the surface $\mathbf{n} \cdot \mathbf{x} = 0$ of the form (1.57) which pertain to a special class of 2-dimensional deformations of the form (1.45), (1.48) or (1.55) and (ii) the tractions needed to sustain them at that surface. As we shall see soon, $\mathbf{Z}$ is a function of $\mathbf{m} \times \mathbf{n}$.

From (1.73) and (1.74) we immediately obtain the integral representation of $\mathbf{Z}$.

Theorem 1.18

$$\mathbf{Z} = \mathbf{S}_2^{-1} + \sqrt{-1}\,\mathbf{S}_2^{-1}\mathbf{S}_1, \tag{1.76}$$

where the matrices $\mathbf{S}_1$ *and* $\mathbf{S}_2$ *are given by (1.68).*

From (1.31), (1.32), (1.68) and the 2π-periodicity of $\cos\phi$ and $\sin\phi$, we observe

Corollary 1.19 *The matrices* $\mathbf{S}_1, \mathbf{S}_2$, *and thence* $\mathbf{Z}$, *are invariant under the rotations of* $\mathbf{m}$ *and* $\mathbf{n}$ *around the vector product* $\mathbf{m} \times \mathbf{n}$. *Hence they are functions not of* $\mathbf{m}$ *and* $\mathbf{n}$ *individually but of the vector product* $\mathbf{m} \times \mathbf{n}$.

This rotational invariance can also be proved directly from (1.74) and Theorems 1.5, 1.9 and 1.10 (Exercise 1-8).

Remark 1.20 There is the arbitrariness of the choice of linearly independent eigenvectors and generalized eigenvectors. However, Theorem 1.18 implies that the arbitrariness is cancelled out in making the product of the two matrices in (1.74), and hence $\mathbf{Z}$ is well-defined.

Theorem 1.21 *The surface impedance tensor* $\mathbf{Z}$ *is Hermitian (i.e.,* $\mathbf{Z} = \overline{\mathbf{Z}}^T$ *) and positive definite.*

Here we prove that $\mathbf{Z}$ is Hermitian. The positive definiteness of $\mathbf{Z}$ is proved in Section 7.D of [22] and [41] indirectly by using the Lagrangian. The direct proof, which is long, is given in [24, 76] and Section 6.6 of [77].

Lemma 1.22 *Let* $\mathbf{S}$ *be the angular average of* $\mathbf{N}(\phi)$ *given in* (1.67). *Then*

$$\mathbf{S}^2 = -\mathbf{I}, \tag{1.77}$$

where $\mathbf{I}$ *denotes the* 6×6 *identity matrix.*

Proof Let $\begin{bmatrix} \mathbf{a}_\alpha \\ \mathbf{l}_\alpha \end{bmatrix}$ $(\alpha = 1, 2, 3)$ be linearly independent eigenvectors or generalized eigenvectors of $\mathbf{N}(0)$ corresponding to the eigenvalues p_α $(\alpha = 1, 2, 3)$ with Im $p_\alpha > 0$. Then from Theorem 1.15 it follows that

$$\mathbf{S}^2 \begin{bmatrix} \mathbf{a}_\alpha \\ \mathbf{l}_\alpha \end{bmatrix} = - \begin{bmatrix} \mathbf{a}_\alpha \\ \mathbf{l}_\alpha \end{bmatrix} \qquad (\alpha = 1, 2, 3). \tag{1.78}$$

Recalling (1.44), we take the complex conjugate of both sides of (1.78) to see that (1.78) holds for six linearly independent eigenvectors or generalized eigenvectors of $\mathbf{N}(0)$. This proves the lemma. □

Proof that $\mathbf{Z}$ *is Hermitian* By the expression of $\mathbf{S}$ in (1.67), the $(1, 2)$ block-component of the equality (1.77) turns out to be

$$\mathbf{S}_1 \mathbf{S}_2 + \mathbf{S}_2 \mathbf{S}_1^T = \mathbf{O}.$$

Since $\mathbf{S}_2$ is symmetric and invertible, it follows that

$$\mathbf{S}_2^{-1} \mathbf{S}_1 = - \left(\mathbf{S}_2^{-1} \mathbf{S}_1 \right)^T,$$

which implies that $\mathbf{S}_2^{-1} \mathbf{S}_1$ is anti-symmetric. Hence $\mathbf{Z}$ in (1.76) is Hermitian. □

Finally, we give one remark.

Remark 1.23 Suppose that $\mathbf{C} = \left(C_{ijkl} \right)_{i,j,k,l=1,2,3}$ enjoys the major symmetry (1.6) but does not have the minor symmetries (1.4). We assume, instead of (1.7), the strong ellipticity condition:

$$\text{The matrix } \left(\sum_{j,l=1}^{3} C_{ijkl} \, \xi_j \bar{\xi}_l \right)_{i\downarrow k \to 1,2,3} \text{ is positive definite for}$$

$$\text{any non-zero vector } \boldsymbol{\xi} = (\xi_1, \xi_2, \xi_3) \in \mathbb{R}^3, \tag{1.79}$$

and assume that the "traction" on $\mathbf{n} \cdot \mathbf{x} = 0$ is given by (1.24). Then all the arguments in Section 1.2 to Section 1.7 except the positive definiteness of $\mathbf{Z}$ remain valid.[12]

Note When the elastic body in question carries residual stress, the elasticity tensor $\mathbf{C}$ in equation (1.9) is replaced by a fourth-order tensor $\mathbf{B}$, whose components, sometimes known [74] as "effective elastic coefficients" in the literature, have the major symmetry but not the minor symmetries [34, 46, 47].

1.8 Examples

We give the formulas for the surface impedance tensors for isotropic and for transversely isotropic media. For this purpose, we first investigate the conditions on $\mathbf{C} = \left(C_{ijkl} \right)_{i,j,k,l=1,2,3}$ and on the vector product $\mathbf{m} \times \mathbf{n}$ for the corresponding Stroh's eigenvalue problem (1.29) to be simple, semisimple, and degenerate (cf. Corollary 1.11). Then we compute the displacement parts and the traction parts of Stroh eigenvectors and generalized eigenvectors. The explicit formulas for the surface impedance tensors obtained here will be used in Chapter 2.

1.8.1 Isotropic Media

For isotropic elasticity, the eigenvalue problem (1.29) is degenerate with an eigenvalue p_1 of multiplicity three and belongs to Case D1 in Section 1.4 for any orthogonal unit vectors $\mathbf{m}$ and $\mathbf{n}$ in $\mathbb{R}^3$. We shall observe this in what follows.

For the orthogonal unit vectors $\mathbf{m} = (m_1, m_2, m_3)$ and $\mathbf{n} = (n_1, n_2, n_3)$, put

$$y_i = m_i + p\, n_i, \qquad i = 1, 2, 3. \tag{1.80}$$

Recalling that the components of the elasticity tensor is written as (see (1.11))

$$C_{ijkl} = \lambda\, \delta_{ij}\delta_{kl} + \mu(\delta_{ik}\delta_{jl} + \delta_{il}\delta_{kj}), \tag{1.81}$$

we see from (1.19) that

$$
\mathbf{Q} + p\left(\mathbf{R} + \mathbf{R}^T\right) + p^2\mathbf{T} = \left(\sum_{j,l=1}^{3} C_{ijkl}\,(m_j + pn_j)(m_l + pn_l) \right)_{i\downarrow k \to 1,2,3}
$$

$$
= \left(\sum_{j,l=1}^{3} C_{ijkl}\, y_j\, y_l \right)_{i\downarrow k \to 1,2,3}
$$

$$
= \left((\lambda + \mu) y_i\, y_k + \mu\, \delta_{ik}\left(y_1^2 + y_2^2 + y_3^2 \right) \right)_{i\downarrow k \to 1,2,3}
$$

$$
= \mu\left(y_1^2 + y_2^2 + y_3^2 \right)\mathbf{I} + (\lambda + \mu)\, (y_i\, y_k)_{i\downarrow k \to 1,2,3}, \tag{1.82}
$$

---

[12]For the positive definiteness of $\mathbf{Z}$, the strong convexity condition (1.7) is needed (see the references in the comment after Theorem 1.21).

where $\mathbf{I}$ is the 3×3 matrix. Then we observe that

$$\det\left[\mathbf{Q} + p\left(\mathbf{R} + \mathbf{R}^T\right) + p^2\mathbf{T}\right] = \mu^2(\lambda + 2\mu)\left(y_1^2 + y_2^2 + y_3^2\right)^3.\text{[13]}$$

Note that the strong convexity condition (1.7) for isotropic elasticity is

$$\mu > 0, \qquad 3\lambda + 2\mu > 0 \tag{1.83}$$

(Exercise 1-2). Hence $\lambda + 2\mu > 0$. Since $\mathbf{m}$ and $\mathbf{n}$ are unit and orthogonal, we have

$$y_1^2 + y_2^2 + y_3^2 = 1 + p^2. \tag{1.84}$$

Hence the sextic equation (1.22) becomes

$$\det\left[\mathbf{Q} + p\left(\mathbf{R} + \mathbf{R}^T\right) + p^2\mathbf{T}\right] = \mu^2(\lambda + 2\mu)\left(1 + p^2\right)^3 = 0,$$

which gives the triple characteristic root with a positive imaginary part

$$p = \sqrt{-1}.$$

This is surely one of the solutions (the singular solution) to the Riccati equation (1.34) (see Remark 1.4).

Then by (1.82) and (1.84), the equation for the displacement part $\mathbf{a}$ of a Stroh eigenvector, (1.21), becomes

$$\left[\mathbf{Q} + p\left(\mathbf{R} + \mathbf{R}^T\right) + p^2\mathbf{T}\right]\mathbf{a} = (\lambda + \mu)\begin{bmatrix} y_1^2 & y_1 y_2 & y_1 y_3 \\ y_1 y_2 & y_2^2 & y_2 y_3 \\ y_1 y_3 & y_2 y_3 & y_3^2 \end{bmatrix}\mathbf{a} = \mathbf{0}. \tag{1.85}$$

Write

$$\mathbf{m} \times \mathbf{n} = \boldsymbol{\ell} = \begin{bmatrix} \ell_1 \\ \ell_2 \\ \ell_3 \end{bmatrix}.$$

Then taking account of

$$y_1\ell_1 + y_2\ell_2 + y_3\ell_3 = 0 \tag{1.86}$$

and equation (1.84) with $p = \sqrt{-1}$, we have two linearly independent solutions to (1.85):

$$\mathbf{a}_1 = \boldsymbol{\ell}, \qquad \mathbf{a}_2 = \begin{bmatrix} y_1 \\ y_2 \\ y_3 \end{bmatrix}_{p=\sqrt{-1}} = \mathbf{m} + \sqrt{-1}\,\mathbf{n}. \tag{1.87}$$

It is obvious that the rank of the 3×3 matrix in (1.85) is equal to 1, which implies that there exist exactly two linearly independent displacement parts corresponding

---

[13]For a 3×3 matrix $\mathbf{M}$ and a scalar q,

$$\det(q\,\mathbf{I} + \mathbf{M}) = q^3 + \operatorname{tr}\mathbf{M}\,q^2 + \frac{1}{2}\left((\operatorname{tr}\mathbf{M})^2 - \operatorname{tr}(\mathbf{M}^2)\right)q + \det\mathbf{M},$$

where tr denotes the trace of a square matrix. Use this formula by putting $q = \mu\left(y_1^2 + y_2^2 + y_3^2\right)$ and $\mathbf{M} = (\lambda + \mu)(y_i y_k)_{i\downarrow k \to 1,2,3}$.

🍃 Springer

to $p = \sqrt{-1}$. Hence, by (1.73), Stroh's eigenvalue problem (1.29) for isotropic elasticity has exactly two linearly independent eigenvectors corresponding to $p = \sqrt{-1}$. Therefore, for isotropic elasticity, the eigenvalue problem (1.29) is degenerate and belongs to Case D1.

Now we compute the traction parts $\mathbf{l}_1$ and $\mathbf{l}_2$ of the Stroh eigenvectors $\begin{bmatrix} \mathbf{a}_\alpha \\ \mathbf{l}_\alpha \end{bmatrix}$ ($\alpha = 1, 2$) where $\mathbf{a}_\alpha$ ($\alpha = 1, 2$) are given by (1.87). By (1.81), (1.25) can be written as

$$[\mathbf{R}^T + p\mathbf{T}]\,\mathbf{a} = \left(\sum_{j,l=1}^{3} C_{ijkl}\, y_l\, n_j\right)_{i\downarrow k \to 1,2,3} \mathbf{a}$$

$$= \left(\lambda\, n_i y_k + \mu(p\, \delta_{ik} + y_i n_k)\right)_{i\downarrow k \to 1,2,3} \mathbf{a}, \tag{1.88}$$

where we have used the relation

$$y_1 n_1 + y_2 n_2 + y_3 n_3 = p. \tag{1.89}$$

Then using the relations (1.84), (1.86), (1.89) with $p = \sqrt{-1}$ and

$$n_1 \ell_1 + n_2 \ell_2 + n_3 \ell_3 = 0,$$

we obtain from (1.87)

$$\mathbf{l}_1 = \mu\, p\, \boldsymbol{\ell} = \sqrt{-1}\,\mu\, \boldsymbol{\ell}, \qquad \mathbf{l}_2 = 2\mu\, p \begin{bmatrix} y_1 \\ y_2 \\ y_3 \end{bmatrix}_{p=\sqrt{-1}} = 2\mu(-\mathbf{n} + \sqrt{-1}\,\mathbf{m}). \tag{1.90}$$

We compute $\mathbf{a}_3$ and $\mathbf{l}_3$ which constitute a generalized eigenvector for (1.29). From (1.87), (1.88), (1.89) and (1.84) with $p = \sqrt{-1}$ we also have

$$[\mathbf{R} + p\mathbf{T}]\,\mathbf{a}_2 = \left(\lambda\, y_i n_k + \mu(p\, \delta_{ik} + n_i y_k)\right)_{i\downarrow k \to 1,2,3} \mathbf{a}_2 = (\lambda + \mu)p \begin{bmatrix} y_1 \\ y_2 \\ y_3 \end{bmatrix}_{p=\sqrt{-1}}.$$

Then by (1.85) and (1.90), equation (1.51) becomes

$$(\lambda + \mu) \begin{bmatrix} y_1^2 & y_1 y_2 & y_1 y_3 \\ y_1 y_2 & y_2^2 & y_2 y_3 \\ y_1 y_3 & y_2 y_3 & y_3^2 \end{bmatrix} \mathbf{a}_3 + (\lambda + 3\mu)p \begin{bmatrix} y_1 \\ y_2 \\ y_3 \end{bmatrix} = \mathbf{0} \quad \left(p = \sqrt{-1}\right). \tag{1.91}$$

Taking account of (1.89), we can solve this for $\mathbf{a}_3$ as

$$\mathbf{a}_3 = -\frac{\lambda + 3\mu}{\lambda + \mu}\,\mathbf{n}. \tag{1.92}$$

It follows from (1.81), (1.87) and (1.89) that

$$\mathbf{T}\,\mathbf{a}_2 = \left(\sum_{j,l=1}^{3} C_{ijkl}\, n_j n_l\right)_{i\downarrow k \to 1,2,3} \mathbf{a}_2 = \left(\lambda\, n_i n_k + \mu\,(\delta_{ik} + n_i n_k)\right)_{i\downarrow k \to 1,2,3} \mathbf{a}_2$$

$$= (\lambda + \mu)p\,\mathbf{n} + \mu \begin{bmatrix} y_1 \\ y_2 \\ y_3 \end{bmatrix} = \mu\,\mathbf{m} + (\lambda + 2\mu)p\,\mathbf{n} \quad (p = \sqrt{-1}).$$

Springer

Hence by (1.88) and (1.92), (1.50) becomes

$$\mathbf{l}_3 = -\frac{\lambda + 3\mu}{\lambda + \mu}\left(\mu\,\mathbf{m} + (\lambda + 2\mu)p\,\mathbf{n}\right) + \mu\,\mathbf{m} + (\lambda + 2\mu)p\,\mathbf{n}$$

$$= \frac{-2\mu}{\lambda + \mu}\left(\mu\,\mathbf{m} + \sqrt{-1}\,(\lambda + 2\mu)\mathbf{n}\right).$$

Using the vectors $\mathbf{a}_\alpha, \mathbf{l}_\alpha\,(\alpha = 1, 2, 3)$ obtained above, we compute the surface impedance tensor for isotropic elasticity. We easily see that

$$\det[\mathbf{a}_1, \mathbf{a}_2, \mathbf{a}_3] = -\frac{\lambda + 3\mu}{\lambda + \mu}$$

and

$$\mathrm{Cof}\,[\mathbf{a}_1, \mathbf{a}_2, \mathbf{a}_3] = \left[-\frac{\lambda + 3\mu}{\lambda + \mu}\,\boldsymbol{\ell},\ -\frac{\lambda + 3\mu}{\lambda + \mu}\,\mathbf{m},\ \mathbf{n} - \sqrt{-1}\,\mathbf{m}\right], \qquad (1.93)$$

where $\mathrm{Cof}\,[\mathbf{a}_1, \mathbf{a}_2, \mathbf{a}_3]$ denotes the cofactor matrix of the matrix $[\mathbf{a}_1, \mathbf{a}_2, \mathbf{a}_3]$. Therefore, by simple algebraic calculations of

$$\mathbf{Z} = -\sqrt{-1}\,[\mathbf{l}_1, \mathbf{l}_2, \mathbf{l}_3][\mathbf{a}_1, \mathbf{a}_2, \mathbf{a}_3]^{-1} = \sqrt{-1}\frac{\lambda + \mu}{\lambda + 3\mu}\,[\mathbf{l}_1, \mathbf{l}_2, \mathbf{l}_3]\,\mathrm{Cof}\,[\mathbf{a}_1, \mathbf{a}_2, \mathbf{a}_3]^T, \quad (1.94)$$

we obtain (Exercise 1-10)

Theorem 1.24 *The surface impedance tensor for isotropic elasticity is given by*

$$\mathbf{Z} = \overline{\mathbf{Z}}^T = \left(Z_{ij}\right)_{i\downarrow j\rightarrow 1,2,3},$$

$$Z_{ii} = \frac{\mu}{\lambda + 3\mu}\left(2(\lambda + 2\mu) - (\lambda + \mu)\,\ell_i^2\right), \quad i = 1, 2, 3,$$

$$Z_{ij} = \frac{\mu}{\lambda + 3\mu}\left(-(\lambda + \mu)\,\ell_i\,\ell_j + \sqrt{-1}(-1)^k\,2\mu\,\ell_k\right), \quad 1 \le i < j \le 3, \quad (1.95)$$

where $(\ell_1, \ell_2, \ell_3) = \mathbf{m} \times \mathbf{n}$ *and the index* k *is determined by* $\{k\} = \{1, 2, 3\} \setminus \{i, j\}$.

To obtain the surface impedance tensor for isotropic elasticity, the method described above is systematic but is not efficient. We give here an alternative method, which takes full advantage of rotational invariance of the surface impedance tensor in the reference plane (Corollary 1.19) and the rotational symmetry of the elasticity tensor itself.

Let us fix the Cartesian coordinate system (x_1, x_2, x_3). Using spherical coordinates in $\mathbb{R}^3$, we write

$$\mathbf{m} \times \mathbf{n} = \boldsymbol{\ell} = \begin{bmatrix} \ell_1 \\ \ell_2 \\ \ell_3 \end{bmatrix} = \begin{bmatrix} \sin\alpha\,\cos\beta \\ \sin\alpha\,\sin\beta \\ \cos\alpha \end{bmatrix} \qquad (0 \le \alpha \le \pi, 0 \le \beta < 2\pi). \qquad (1.96)$$

Then for

$$\boldsymbol{\ell}_0 = \begin{bmatrix} 0 \\ 0 \\ 1 \end{bmatrix}$$

we have

$$\ell = \Omega_3 \, \Omega_2 \, \ell_0,$$

where

$$\Omega_2 = \begin{bmatrix} \cos\alpha & 0 & \sin\alpha \\ 0 & 1 & 0 \\ -\sin\alpha & 0 & \cos\alpha \end{bmatrix}$$

is the transformation matrix pertaining to the rotation of the material around the 2-axis by the angle α and

$$\Omega_3 = \begin{bmatrix} \cos\beta & -\sin\beta & 0 \\ \sin\beta & \cos\beta & 0 \\ 0 & 0 & 1 \end{bmatrix} \tag{1.97}$$

is that pertaining to the rotation of the material around the 3-axis by the angle β. Since the components of the isotropic elasticity tensor (1.81) remain invariant under these transformations, we apply the transformation formula for second order tensors to get[14]

$$\mathbf{Z} = \Omega_3 \, \Omega_2 \, \mathbf{Z}_0 \, (\Omega_3 \, \Omega_2)^T, \tag{1.99}$$

where $\mathbf{Z}_0$ is the surface impedance tensor for $\mathbf{m}$ and $\mathbf{n}$ such that $\mathbf{m} \times \mathbf{n} = \ell_0 = (0, 0, 1)$. Hence, by Corollary 1.19, we may take

$$\mathbf{m} = \mathbf{m}_0 = \begin{bmatrix} 1 \\ 0 \\ 0 \end{bmatrix}, \quad \mathbf{n} = \mathbf{n}_0 = \begin{bmatrix} 0 \\ 1 \\ 0 \end{bmatrix}.$$

Then, we see that (1.85), (1.88) and (1.91) assume more simplified forms, and we easily get the corresponding $\mathbf{a}_\alpha, \mathbf{l}_\alpha \, (\alpha = 1, 2, 3)$ to obtain

$$\mathbf{Z}_0 = \frac{\mu}{\lambda + 3\mu} \begin{bmatrix} 2(\lambda + 2\mu) & -2\sqrt{-1}\,\mu & 0 \\ 2\sqrt{-1}\,\mu & 2(\lambda + 2\mu) & 0 \\ 0 & 0 & \lambda + 3\mu \end{bmatrix}.$$

Finally, from (1.96) and (1.99) we obtain (1.95).

1.8.2 Transversely Isotropic Media

For media that are transversely isotropic, the eigenvalue problem (1.29) becomes either simple, semisimple, or degenerate with the appearance of a Jordan chain

---

[14]Recalling that $\mathbf{Z}$ is a function of $\mathbf{m} \times \mathbf{n} = \ell$ (Corollary 1.19), we write $\mathbf{Z}$ as $\mathbf{Z}(\ell)$. Then the transformation formula for second order tensors implies that for any orthogonal tensor $\mathbf{Q}$

$$\mathbf{Z}^*(\mathbf{Q}\,\ell_0) = \mathbf{Q}\,\mathbf{Z}(\ell_0)\,\mathbf{Q}^T, \tag{1.98}$$

where $\mathbf{Z}^*$ is the surface impedance tensor after the material is rotated by $\mathbf{Q}$.
For an isotropic material, $\mathbf{Z} = \mathbf{Z}^*$, and it follows from (1.98) that

$$\mathbf{Z}(\mathbf{Q}\,\ell_0) = \mathbf{Q}\,\mathbf{Z}(\ell_0)\,\mathbf{Q}^T.$$

Putting $\mathbf{Q} = \Omega_3 \, \Omega_2$, we arrive at (1.99).

of length two. These conditions are determined by the elasticity tensor $\mathbf{C} = (C_{ijkl})_{1 \leq i,j,k,l \leq 3}$ and the vector product $\mathbf{m} \times \mathbf{n}$ of the orthogonal unit vectors $\mathbf{m}$ and $\mathbf{n}$ (Corollary 1.11).

The elasticity tensor of a transversely isotropic medium whose axis of symmetry coincides with the 3-axis is given by (1.13). Hereafter, for simplicity we put

$$C_{1111} = A, \quad C_{1122} = N, \quad C_{1133} = F, \quad C_{3333} = C, \quad C_{2323} = L. \tag{1.100}$$

The strong convexity condition (1.7) is equivalent to (Exercise 1-3)

$$L > 0, \quad \frac{1}{2}(A - N) > 0, \quad A + C + N > 0, \quad (A + N)C > 2F^2. \tag{1.101}$$

As in the previous subsection, we write $\mathbf{m} \times \mathbf{n} = \boldsymbol{\ell} = (\ell_1, \ell_2, \ell_3)$. Then the conditions for the eigenvalue problem (1.29) to be simple, semisimple and degenerate are given, respectively, as follows:

Lemma 1.25

> (1.29) *is degenerate and belongs to Case* D1
>
> $\iff \ell_3 = 1$ *or* $\ell_3 = -1$ *or* $\sqrt{AC} - F - 2L = 0$
>
> (1.29) *is semisimple*
>
> $\iff \ell_3 \neq \pm 1$ *and* $\sqrt{AC} - F - 2L \neq 0$ *and*
>
> $$AL \left(\frac{2L}{A - N} \right)^2 - (AC - F^2 - 2FL)\frac{2L}{A - N} + CL = 0 \tag{1.102}$$
>
> (1.29) *is simple*
>
> $\iff$ *all other conditions.*

Theorem 1.26 *The surface impedance tensor for a transversely isotropic medium whose axis of symmetry coincides with the 3-axis is given by*

$$\mathbf{Z} = (Z_{ij})_{i \downarrow j \to 1,2,3} = \overline{\mathbf{Z}}^T,$$

$$Z_{11} = \frac{L}{K} \left\{ \ell_1^2 + \frac{AG}{D} \left(\ell_3^2 + \frac{2L}{A - N} \ell_2^2 \right) \left(\ell_1^2 + \ell_2^2 \right) \right\},$$

$$Z_{22} = \frac{L}{K} \left\{ \ell_2^2 + \frac{AG}{D} \left(\ell_3^2 + \frac{2L}{A - N} \ell_1^2 \right) \left(\ell_1^2 + \ell_2^2 \right) \right\},$$

$$Z_{33} = \frac{1}{D} \left[AL\left(-\ell_3^4 - H\ell_3^2 + GHK \right) + \{(F + L)^2 - AC\} \left(\ell_1^2 + \ell_2^2 \right)\ell_3^2 \right],$$

⌂ Springer

$$Z_{12} = \frac{L}{K}\left\{1 - \frac{AG}{D}\frac{2L}{A-N}\left(\ell_1{}^2 + \ell_2{}^2\right)\right\}\ell_1\ell_2 - \sqrt{-1}\left\{A - N - \frac{AGL}{D}\left(\ell_1{}^2 + \ell_2{}^2\right)\right\}\ell_3,$$

$$Z_{13} = \frac{-L(F+L)}{D}\left(\ell_1{}^2 + \ell_2{}^2\right)\ell_1\ell_3 - \sqrt{-1}L\left\{-1 + \frac{(F+L)K}{D}\left(\ell_1{}^2 + \ell_2{}^2\right)\right\}\ell_2,$$

$$Z_{23} = \frac{-L(F+L)}{D}\left(\ell_1{}^2 + \ell_2{}^2\right)\ell_2\ell_3 + \sqrt{-1}L\left\{-1 + \frac{(F+L)K}{D}\left(\ell_1{}^2 + \ell_2{}^2\right)\right\}\ell_1,$$

$$(1.103)$$

where

$$K = \sqrt{\ell_3{}^2 + \frac{2L}{A-N}\left(\ell_1{}^2 + \ell_2{}^2\right)},$$

$$G = \sqrt{\frac{2AL\,\ell_3{}^2 + (AC - F^2 - 2FL)\left(\ell_1{}^2 + \ell_2{}^2\right) + 2\sqrt{AL}\sqrt{\Delta}}{AL}},$$

$$H = \sqrt{\frac{\Delta}{AL}}, \qquad D = \left\{A\,\ell_3{}^2 + L\left(\ell_1{}^2 + \ell_2{}^2\right) + AH\right\}K - AG\,\ell_3{}^2,$$

$$\Delta = AL\,\ell_3{}^4 + (AC - F^2 - 2FL)\left(\ell_1{}^2 + \ell_2{}^2\right)\ell_3{}^2 + CL\left(\ell_1{}^2 + \ell_2{}^2\right)^2. \qquad (1.104)$$

Note In the case $\ell_3 = \pm 1$, we have $\ell_1{}^2 + \ell_2{}^2 = 0$. Then $K = 1, \Delta = AL, G = 2,$ $H = 1$ and $D = 0$. Hence each Z_{ij} in (1.103) has a indeterminate form. In this case, $\mathbf{Z}$ is obtained by taking the limit $\ell_1{}^2 + \ell_2{}^2 \longrightarrow 0$ in (1.103). In fact, as $\ell_1{}^2 + \ell_2{}^2 \longrightarrow 0$ we get

$$K = 1 + \frac{2L - A + N}{2(A - N)}\left(\ell_1{}^2 + \ell_2{}^2\right) + O\left(\left(\ell_1{}^2 + \ell_2{}^2\right)^2\right),$$

$$\Delta = AL + (AC - F^2 - 2FL - 2AL)\left(\ell_1{}^2 + \ell_2{}^2\right) + O\left(\left(\ell_1{}^2 + \ell_2{}^2\right)^2\right)$$

and

$$G = 2 + \frac{AC - F^2 - 2FL - 2AL}{2AL}\left(\ell_1{}^2 + \ell_2{}^2\right) + O\left(\left(\ell_1{}^2 + \ell_2{}^2\right)^2\right),$$

$$H = 1 + \frac{AC - F^2 - 2FL - 2AL}{2AL}\left(\ell_1{}^2 + \ell_2{}^2\right) + O\left(\left(\ell_1{}^2 + \ell_2{}^2\right)^2\right).$$

Then

$$D = \frac{L(3A - N)}{A - N}\left(\ell_1{}^2 + \ell_2{}^2\right) + O\left(\left(\ell_1{}^2 + \ell_2{}^2\right)^2\right).$$

Substituting these into (1.103), we let $\ell_1{}^2 + \ell_2{}^2 \longrightarrow 0$. Then the components of $\mathbf{Z}$ in the case $\ell_3 = \pm 1$ become

$$Z_{11} = Z_{22} = \frac{2A(A - N)}{3A - N}, \qquad Z_{33} = L,$$

$$Z_{12} = -\sqrt{-1}\,\frac{(A - N)^2}{3A - N}\ell_3, \qquad Z_{13} = Z_{23} = 0. \qquad (1.105)$$

We note that $\ell_3 = \pm 1$ implies that the reference plane (i.e., the plane generated by the orthogonal unit vectors $\mathbf{m}$ and $\mathbf{n}$) is perpendicular to the axis of symmetry of the transversely isotropic medium in question.

Proof of Lemma 1.25 and Theorem 1.26 As in (1.96), we write

$$\mathbf{m} \times \mathbf{n} = \boldsymbol{\ell} = \begin{bmatrix} \ell_1 \\ \ell_2 \\ \ell_3 \end{bmatrix} = \begin{bmatrix} \sin\alpha \cos\beta \\ \sin\alpha \sin\beta \\ \cos\alpha \end{bmatrix} \quad (0 \le \alpha \le \pi, 0 \le \beta < 2\pi). \tag{1.106}$$

Then for

$$\boldsymbol{\ell}_0 = \begin{bmatrix} \sin\alpha \\ 0 \\ \cos\alpha \end{bmatrix}$$

we have

$$\boldsymbol{\ell} = \Omega_3 \boldsymbol{\ell}_0,$$

where Ω_3 is the transformation matrix (1.97) related to the rotation of the material around the 3-axis by the angle β. Since the 3-axis has been taken to be the axis of symmetry, the components of the transversely isotropic elasticity tensor (1.100) remain invariant under the transformation by Ω_3. Hence the transformation formula for the second order tensor can be applied to get[15]

$$\mathbf{Z} = \Omega_3 \, \mathbf{Z}_0 \, \Omega_3^T, \tag{1.107}$$

where $\mathbf{Z}_0$ is the surface impedance tensor for $\mathbf{m}$ and $\mathbf{n}$ such that $\mathbf{m} \times \mathbf{n} = \boldsymbol{\ell}_0 = (\sin\alpha, 0, \cos\alpha)$. Then, by Corollary 1.19, we may take

$$\mathbf{m} = \mathbf{m}_0 = \begin{bmatrix} \cos\alpha \\ 0 \\ -\sin\alpha \end{bmatrix}, \qquad \mathbf{n} = \mathbf{n}_0 = \begin{bmatrix} 0 \\ 1 \\ 0 \end{bmatrix}. \tag{1.108}$$

From Corollary 1.11 it is also possible to consider degeneracy of the eigenvalue problem (1.29) by using (1.108).

Let y_i ($i = 1, 2, 3$) be given by (1.80). By (1.100), equation (1.21) becomes

$$[\mathbf{Q} + p\,(\mathbf{R} + \mathbf{R}^T) + p^2\mathbf{T}]\,\mathbf{a} = \left(\sum_{j,l=1}^{3} C_{ijkl}\, y_j\, y_l \right)_{i\downarrow k \to 1,2,3} \mathbf{a}$$

$$= \begin{bmatrix} Ay_1^2 + \frac{1}{2}(A-N)y_2^2 + Ly_3^2 & \frac{1}{2}(A+N)y_1 y_2 & (F+L)y_1 y_3 \\ & \frac{1}{2}(A-N)y_1^2 + Ay_2^2 + Ly_3^2 & (F+L)y_2 y_3 \\ \text{Sym.} & & L(y_1^2 + y_2^2) + Cy_3^2 \end{bmatrix} \mathbf{a}$$

$$= \mathbf{0}. \tag{1.109}$$

---

[15] Again we use (1.98). For a transversely isotropic material in question, $\mathbf{Z} = \mathbf{Z}^*$ for $\mathbf{Q} = \Omega_3$. Then

$$\mathbf{Z}(\mathbf{Q}\,\boldsymbol{\ell}_0) = \mathbf{Q}\,\mathbf{Z}(\boldsymbol{\ell}_0)\,\mathbf{Q}^T$$

for $\mathbf{Q} = \Omega_3$, which implies (1.107).

The sextic equation (1.22) can be factorized as

$$\det\left[\mathbf{Q} + p\left(\mathbf{R} + \mathbf{R}^T\right) + p^2\mathbf{T}\right] = \left(\frac{1}{2}(A - N)(y_1^2 + y_2^2) + Ly_3^2\right)$$
$$\times\left(AL\left(y_1^2 + y_2^2\right)^2 + (AC - F^2 - 2FL)\right)$$
$$\times\left(y_1^2 + y_2^2\right)y_3^2 + CLy_3^4\right) = 0.[16]$$

Since (1.108) implies that

$$y_1 = \cos\alpha, \qquad y_2 = p, \qquad y_3 = -\sin\alpha, \qquad (1.110)$$

we let p_1 be the root with $\operatorname{Im} p_1 > 0$ of the quadratic equation

$$\frac{1}{2}(A - N)\left(y_1^2 + y_2^2\right) + Ly_3^2 = \frac{1}{2}(A - N)\left(p^2 + \cos^2\alpha\right) + L\sin^2\alpha = 0 \qquad (1.111)$$

and p_2, p_3 be the roots with $\operatorname{Im} p_\alpha > 0$ ($\alpha = 2, 3$) of the quartic equation

$$AL\left(y_1^2 + y_2^2\right)^2 + (AC - F^2 - 2FL)\left(y_1^2 + y_2^2\right)y_3^2 + CLy_3^4 = 0. \qquad (1.112)$$

Suppose that $F + L \neq 0$ and $p_2 \neq p_3$. Then we can check from (1.111) and (1.112) that three vectors

$$\mathbf{a}_1 = \begin{bmatrix} y_2 \\ -y_1 \\ 0 \end{bmatrix}_{p=p_1}, \qquad \mathbf{a}_\alpha = \begin{bmatrix} (F+L)y_1y_3 \\ (F+L)y_2y_3 \\ -\left\{A(y_1^2 + y_2^2) + Ly_3^2\right\} \end{bmatrix}_{p=p_\alpha} \qquad (\alpha = 2, 3) \qquad (1.113)$$

satisfy (1.109). From (1.112) it follows that $A\left(y_1^2 + y_2^2\right) + Ly_3^2 \neq 0$ at $p = p_2$ and $p = p_3$, since otherwise, (1.112) becomes $\frac{L}{A}(F + L)^2 y_3^4 = 0$. From the supposition and (1.101) we have $F + L \neq 0$, $A > 0$, $L > 0$. When $y_3 = 0$, (1.112) becomes $\left(y_1^2 + y_2^2\right)^2 = 0$, which contradicts $p_2 \neq p_3$. Therefore, $\mathbf{a}_1$ and $\mathbf{a}_2$ or $\mathbf{a}_1$ and $\mathbf{a}_3$ are linearly independent even if $p_1 = p_2$ or $p_1 = p_3$. Then from (1.73) we see that there exist three linearly independent eigenvectors of the eigenvalue problem (1.29) associated with the eigenvalues p_α ($\alpha = 1, 2, 3$). Hence the eigenvalue problem (1.29) is simple or semisimple.

By (1.100) and (1.108), (1.25) can be written as

$$\left[\mathbf{R}^T + p\mathbf{T}\right]\mathbf{a} = \left(\sum_{j,l=1}^{3} C_{ijkl}\, y_l\, n_j\right)_{i\downarrow k\to 1,2,3} \mathbf{a}$$
$$= \begin{bmatrix} \frac{1}{2}(A - N)y_2 & \frac{1}{2}(A - N)y_1 & 0 \\ Ny_1 & Ay_2 & Fy_3 \\ 0 & Ly_3 & Ly_2 \end{bmatrix}\mathbf{a}.$$

---

[16]This factorization is classically well known [73].

Then from (1.113) we obtain

$$
\mathbf{l}_1 = \begin{bmatrix} \frac{1}{2}(A - N)(y_2^2 - y_1^2) \\ (N - A)y_1 y_2 \\ -L y_1 y_3 \end{bmatrix}_{p=p_1},
$$

$$
\mathbf{l}_\alpha = \begin{bmatrix} (A - N)(F + L)y_1 y_2 y_3 \\ \left\{ A L y_2^2 + ((F + L)N - AF)y_1^2 - F L y_3^2 \right\} y_3 \\ \left\{ -A(y_1^2 + y_2^2) + F y_3^2 \right\} L y_2 \end{bmatrix}_{p=p_\alpha} \qquad (\alpha = 2, 3). \quad (1.114)
$$

On the other hand, we easily see from (1.110) and (1.113) that

$$
\det [\mathbf{a}_1, \mathbf{a}_2, \mathbf{a}_3] = (F + L)(p_2 - p_3)\sin\alpha
$$
$$
\times \left((A\cos^2\alpha + L\sin^2\alpha - A p_2 p_3) p_1 - A(p_2 + p_3)\cos^2\alpha \right)
$$

$$(1.115)$$

and

$$
\text{Cof } [\mathbf{a}_1, \mathbf{a}_2, \mathbf{a}_3] = \left(A_{ij} \right)_{i\downarrow j \to 1,2,3},
$$

where

$$
A_{11} = (F + L)(p_2 - p_3)\sin\alpha \left(A\cos^2\alpha + L\sin^2\alpha - A p_2 p_3 \right),
$$
$$
A_{12} = -\left\{ A\left(\cos^2\alpha + p_3^2 \right) + L\sin^2\alpha \right\} \cos\alpha,
$$
$$
A_{13} = \left\{ A\left(\cos^2\alpha + p_2^2 \right) + L\sin^2\alpha \right\} \cos\alpha,
$$
$$
A_{21} = -A(F + L)\left(p_3^2 - p_2^2 \right)\cos\alpha\sin\alpha,
$$
$$
A_{22} = -\left\{ A\left(\cos^2\alpha + p_3^2 \right) + L\sin^2\alpha \right\} p_1,
$$
$$
A_{23} = \left\{ A\left(\cos^2\alpha + p_2^2 \right) + L\sin^2\alpha \right\} p_1,
$$
$$
A_{31} = (F + L)^2\left(p_3 - p_2 \right)\cos\alpha\sin^2\alpha,
$$
$$
A_{32} = (F + L)\left(p_1 p_3 + \cos^2\alpha \right)\sin\alpha,
$$
$$
A_{33} = -(F + L)\left(p_1 p_2 + \cos^2\alpha \right)\sin\alpha. \quad (1.116)
$$

In the computations of

$$
\mathbf{Z}_0 = \left(Z_{ij}^0 \right)_{i\downarrow j \to 1,2,3} = -\sqrt{-1}\, [\mathbf{l}_1, \mathbf{l}_2, \mathbf{l}_3][\mathbf{a}_1, \mathbf{a}_2, \mathbf{a}_3]^{-1}
$$
$$
= -\sqrt{-1}\, \frac{1}{\det [\mathbf{a}_1, \mathbf{a}_2, \mathbf{a}_3]}\, [\mathbf{l}_1, \mathbf{l}_2, \mathbf{l}_3]\, \text{Cof } [\mathbf{a}_1, \mathbf{a}_2, \mathbf{a}_3]^T,
$$

we need to express p_1, $p_2 + p_3$ and $p_2 p_3$ in terms of the elastic tensor (1.100). For p_1, we use (1.111) to get

$$p_1 = \sqrt{-1}\sqrt{\cos^2 \alpha + \frac{2L}{A - N} \sin^2 \alpha} = \sqrt{-1}\, K.$$

For p_α ($\alpha = 2, 3$), we rewrite (1.112) using (1.110) as

$$AL p^4 + \left\{ 2AL \cos^2 \alpha + (AC - F^2 - 2FL) \sin^2 \alpha \right\} p^2$$
$$+ AL \cos^4 \alpha + (AC - F^2 - 2FL) \cos^2 \alpha \sin^2 \alpha + CL \sin^4 \alpha = 0,$$

which, from the relations between the roots p_2^2, p_3^2 and the coefficients of the equation above, implies that

$$p_2^2 + p_3^2 = -\frac{2AL \cos^2 \alpha + (AC - F^2 - 2FL) \sin^2 \alpha}{AL},$$

$$p_2^2 p_3^2 = \frac{AL \cos^4 \alpha + (AC - F^2 - 2FL) \cos^2 \alpha \sin^2 \alpha + CL \sin^4 \alpha}{AL} = \frac{\Delta}{AL}.$$

$$(1.117)$$

The numerator Δ can be written as

$$L(\sqrt{A} \cos^2 \alpha - \sqrt{C} \sin^2 \alpha)^2 + (\sqrt{AC} - F)(\sqrt{AC} + F + 2L) \cos^2 \alpha \sin^2 \alpha,$$

which is positive, because the strong convexity condition (1.101) implies that $AC - F^2 > 0$, $\sqrt{AC} - F > 0$ and $\sqrt{AC} + F > 0$. Taking account of Im $p_\alpha > 0$ ($\alpha = 2, 3$), we have

$$p_2 p_3 = -\sqrt{\frac{\Delta}{AL}} = -H.$$

Then from (1.117) it follows that

$$(p_2 + p_3)^2 = p_2^2 + p_3^2 + 2 p_2 p_3$$
$$= -\frac{2AL \cos^2 \alpha + (AC - F^2 - 2FL) \sin^2 \alpha + 2\sqrt{AL}\sqrt{\Delta}}{AL},$$

and since $\mathrm{Im}(p_2 + p_3) > 0$, we have

$$p_2 + p_3 = \sqrt{-1}\sqrt{\frac{2AL \cos^2 \alpha + (AC - F^2 - 2FL) \sin^2 \alpha + 2\sqrt{AL}\sqrt{\Delta}}{AL}}$$
$$= \sqrt{-1}\, G.$$

Let us compute Z^0_{11}. Denoting the (i, j) entry of the matrix $[\mathbf{l}_1, \mathbf{l}_2, \mathbf{l}_3]$ by L_{ij}, we see from (1.114) and (1.116) that the $(1, 1)$ entry of the matrix $[\mathbf{l}_1, \mathbf{l}_2, \mathbf{l}_3]\,\mathrm{Cof}[\mathbf{a}_1, \mathbf{a}_2, \mathbf{a}_3]^T$ is

$$L_{11}A_{11} + L_{12}A_{12} + L_{13}A_{13}$$

$$= (A - N)(F + L)\,(p_2 - p_3)\sin\alpha \left\{ \frac{p_1^2 - \cos^2\alpha}{2}\left(A\cos^2\alpha + L\sin^2\alpha - Ap_2p_3\right)\right.$$

$$\left. + \left(A\cos^2\alpha + L\sin^2\alpha\right)\cos^2\alpha - Ap_2p_3\cos^2\alpha\right\}$$

$$= (A - N)(F + L)(p_2 - p_3)\sin\alpha$$

$$\times \left\{ \left(-\cos^2\alpha - \frac{L}{A - N}\sin^2\alpha\right)(A\cos^2\alpha + L\sin^2\alpha + AH)\right.$$

$$\left. + \left(A\cos^2\alpha + L\sin^2\alpha\right)\cos^2\alpha + AH\cos^2\alpha\right\}$$

$$= -(F + L)L(p_2 - p_3)\left(A\cos^2\alpha + L\sin^2\alpha + AH\right)\sin^3\alpha.$$

On the other hand, (1.115) can be written as

$$\det[\mathbf{a}_1, \mathbf{a}_2, \mathbf{a}_3] = \sqrt{-1}\,(F + L)(p_2 - p_3)\sin\alpha$$

$$\times \left((A\cos^2\alpha + L\sin^2\alpha + AH)K - AG\cos^2\alpha\right)$$

$$= \sqrt{-1}\,(F + L)(p_2 - p_3)D\sin\alpha.$$

Hence

$$Z^0_{11} = \frac{L\sin^2\alpha(A\cos^2\alpha + L\sin^2\alpha + AH)}{D} = \frac{L\sin^2\alpha(D + AG\cos^2\alpha)}{KD}.$$

In the same way, by simple but long algebraic calculations, we obtain the other entries:

$$\mathbf{Z}_0 = \left(Z^0_{ij}\right)_{i\downarrow j \to 1,2,3} = \overline{\mathbf{Z}_0}^T,$$

$$Z^0_{11} = \frac{L}{K}\sin^2\alpha\left(1 + \frac{AG}{D}\cos^2\alpha\right), \quad Z^0_{22} = \frac{LAGK}{D}\sin^2\alpha,$$

$$Z^0_{33} = \frac{1}{D}\left[AL(-\cos^4\alpha - H\cos^2\alpha + GHK)\right.$$

$$\left. + \cos^2\alpha\sin^2\alpha\left\{(F + L)^2 - AC\right\}\right],$$

$$Z^0_{12} = -\sqrt{-1}\,\cos\alpha\left(A - N - \frac{AGL}{D}\sin^2\alpha\right),$$

$$Z^0_{13} = -\frac{L(F + L)}{D}\cos\alpha\sin^3\alpha,$$

$$Z^0_{23} = \sqrt{-1}\,L\sin\alpha\left(-1 + \frac{(F + L)K}{D}\sin^2\alpha\right). \tag{1.118}$$

Springer

Then from (1.106) and (1.107) we obtain (1.103). This is the generic case. Henceforth, we consider the missing cases, i.e., the case where $p_2 = p_3$ and the case where $F + L = 0$, and we give an outline of the computations. For the details we refer to [69].

First, we consider the case $p_2 = p_3$. By virtue of (1.110), equation (1.112) becomes a quadratic equation in p^2:

$$\left(y_1^2 + y_2^2 + Iy_3^2\right)^2 + Jy_3^4 = \left(p^2 + \cos^2\alpha + I\sin^2\alpha\right)^2 + J\sin^4\alpha = 0, \tag{1.119}$$

where

$$I = \frac{AC - F^2 - 2FL}{2AL},$$

$$J = \frac{-(AC - F^2)(\sqrt{AC} + F + 2L)(\sqrt{AC} - F - 2L)}{4A^2L^2}.$$

Hence $p_2 = p_3$ is equivalent to

$$J\sin^4\alpha = 0,$$

which, by (1.101) and $\ell_3 = \cos\alpha$, is also equivalent to

$$\ell_3 = 1 \quad \text{or} \quad \ell_3 = -1 \quad \text{or} \quad \sqrt{AC} - F - 2L = 0. \tag{1.120}$$

Suppose that $\ell_3 = 1$ or $\ell_3 = -1$. Then we have $\sin\alpha = 0$, $\cos^2\alpha = 1$ and $p_1 = p_2 = p_3 = \sqrt{-1}$. From (1.109) and (1.110) we can take two linearly independent vectors

$$\mathbf{a}_1 = \begin{bmatrix} 0 \\ 0 \\ 1 \end{bmatrix}, \qquad \mathbf{a}_2 = \begin{bmatrix} \ell_3 \\ \sqrt{-1} \\ 0 \end{bmatrix}$$

and we can observe that the rank of the 3×3 matrix in (1.109) is one at $p = \sqrt{-1}$. Hence by (1.73), Stroh's eigenvalue problem (1.29) has exactly two linearly independent eigenvectors associated with the eigenvalue $p = \sqrt{-1}$ of multiplicity 3. Thus, the eigenvalue problem (1.29) is degenerate and belongs to Case D1. Then from (1.25), (1.50), (1.51) and (1.74) we obtain

$$Z_{11}^0 = Z_{22}^0 = \frac{2A(A - N)}{3A - N}, \qquad Z_{33}^0 = L,$$

$$Z_{12}^0 = -\sqrt{-1}\,\frac{(A - N)^2}{3A - N}\,\ell_3 = -Z_{21}^0, \qquad Z_{13}^0 = Z_{23}^0 = Z_{31}^0 = Z_{32}^0 = 0.$$

We can check that taking the limit $\sin^2\alpha \longrightarrow 0$ in (1.118) gives the above Z_{ij}^0. Thus we obtain (1.105) from (1.107).

Suppose that $\ell_3 \neq \pm 1$ and $\sqrt{AC} - F - 2L = 0$. Then we get two vectors

$$\mathbf{a}_1 = \begin{bmatrix} y_2 \\ -y_1 \\ 0 \end{bmatrix}_{p=p_1}, \qquad \mathbf{a}_2 = \begin{bmatrix} y_1 \\ y_2 \\ y_3 \end{bmatrix}_{p=p_2},$$

where p_1 ($\operatorname{Im} p_1 > 0$) is the root of (1.111) and $p_2 = p_3$ ($\operatorname{Im} p_2 = \operatorname{Im} p_3 > 0$) is the root of

$$p^2 + \cos^2\alpha + \sqrt{\frac{C}{A}}\sin^2\alpha = 0.$$

We can observe that the rank of the 3×3 matrix in (1.109) is two at $p = p_2$ when $p_1 \neq p_2$ and is one at $p = p_1$ when $p_1 = p_2$. Hence by (1.73), Stroh's eigenvalue problem (1.29) has exactly one eigenvector associated with the eigenvalue p_2 of multiplicity 2 when $p_1 \neq p_2$ and exactly two linearly independent eigenvectors associated with the eigenvalue p_1 of multiplicity 3 when $p_1 = p_2$. Thus, the eigenvalue problem (1.29) is degenerate and belongs to Case D1. Using (1.25), (1.50), (1.51) and (1.74), we can observe that formula (1.118) applies to the present case.

Now let $p_2 \neq p_3$ and $F + L = 0$. Then instead of $\mathbf{a}_\alpha$ $(\alpha = 2, 3)$ in (1.113) we can take

$$
\mathbf{a}_2 = \begin{bmatrix} 0 \\ 0 \\ 1 \end{bmatrix}, \quad
\mathbf{a}_3 = \begin{bmatrix} y_1 \\ y_2 \\ 0 \end{bmatrix}_{p=p_3} ,
$$

where p_2 $(\operatorname{Im} p_2 > 0)$ is the root of

$$
L \left(p^2 + \cos^2 \alpha \right) + C \sin^2 \alpha = 0
$$

and p_3 $(\operatorname{Im} p_3 > 0)$ is the root of

$$
A \left(p^2 + \cos^2 \alpha \right) + L \sin^2 \alpha = 0.
$$

Since the strong convexity condition (1.101) implies that $\frac{A-N}{2} \neq A$, it follows that $p_1 \neq p_3$. Obviously, $\mathbf{a}_1$ in (1.113) and $\mathbf{a}_2$ above are linearly independent even when $p_1 = p_2$. Therefore, the eigenvalue problem (1.29) is simple or semisimple. Then using (1.25) and (1.74), we can observe that formula (1.118) also applies to the present case.

We note that equality (1.102) holds if and only if (1.111) and (1.112) have a common root, that is, $p_1 = p_2$ or $p_1 = p_3$. This makes the eigenvalue problem (1.29) semisimple with an eigenvalue of multiplicity 2, or degenerate with an eigenvalue of multiplicity 3, but the length of the Jordan chain remains the same as before. $\square$

Note The eigenvalue problem (1.29) becomes degenerate with an eigenvalue of multiplicity 3 if and only if $\ell_3 = 1$ or $\ell_3 = -1$ or

$$
\sqrt{AC} - F - 2L = 0 \quad \text{and} \quad 2L/(A - N) = \sqrt{C/A}.
$$

The second equality above follows from $\sqrt{AC} - F - 2L = 0$ and (1.102). Isotropic elasticity belongs to this class.

1.9 Justification of the Solutions in the Stroh Formalism

We have begun our presentation of the Stroh formalism by seeking solutions of the form (1.18) and have called their superposition (1.45) and its modifications (1.48) and (1.55) the general forms of the solution to (1.17) which describes two-dimensional[17] deformations in an elastic half-space $\mathbf{n} \cdot \mathbf{x} \le 0$, satisfies the condition $u = \mathbf{a} e^{-\sqrt{-1}\xi \, \mathbf{m} \cdot \mathbf{x}}$ on $\mathbf{n} \cdot \mathbf{x} = 0$ for some complex $\mathbf{a} \in \mathbb{C}^3$, and decays to zero as $\mathbf{n} \cdot \mathbf{x} \longrightarrow -\infty$. In this section, we justify these forms.

---

[17]For the meaning of "two-dimensional", see the third paragraph of Section 1.2.

❧ Springer

We introduce the variables

$$y_1 = \mathbf{m} \cdot \mathbf{x} = m_1 x_1 + m_2 x_2 + m_3 x_3, \quad y_2 = \mathbf{n} \cdot \mathbf{x} = n_1 x_1 + n_2 x_2 + n_3 x_3. \quad (1.121)$$

Then the solutions considered are defined in $y_2 \leq 0$ and depends only on y_1 and y_2. Hence, noting that

$$\frac{\partial}{\partial x_j} = m_j \frac{\partial}{\partial y_1} + n_j \frac{\partial}{\partial y_2},$$

we can rewrite (1.17) as

$$\left[\mathbf{Q} \frac{\partial^2}{\partial y_1^2} + (\mathbf{R} + \mathbf{R}^T) \frac{\partial^2}{\partial y_1 \partial y_2} + \mathbf{T} \frac{\partial^2}{\partial y_2^2} \right] \mathbf{u}(y_1, y_2) = \mathbf{0} \quad \text{in} \quad y_2 \leq 0, \quad (1.122)$$

where $\mathbf{Q}$, $\mathbf{R}$ and $\mathbf{T}$ are matrices defined by (1.20).

We impose the boundary condition at $y_2 = 0$

$$\mathbf{u}(y_1, 0) = \mathbf{f}(y_1), \quad (1.123)$$

where $\mathbf{f}(y_1)$ is some prescribed boundary data valued in $\mathbb{C}^3$.

Applying the Fourier transform with respect to y_1 to the boundary-value problem (1.122) and (1.123), we obtain

$$\left[(\sqrt{-1}\xi_1)^2 \mathbf{Q} + (\sqrt{-1}\xi_1)(\mathbf{R} + \mathbf{R}^T) \frac{\partial}{\partial y_2} + \mathbf{T} \frac{\partial^2}{\partial y_2^2} \right] \widehat{\mathbf{u}}(\xi_1, y_2) = \mathbf{0} \quad \text{in} \quad y_2 \leq 0 \quad (1.124)$$

and

$$\widehat{\mathbf{u}}(\xi_1, 0) = \widehat{\mathbf{f}}(\xi_1),$$

where $\widehat{}$ denotes the Fourier transform with respect to y_1, i.e.,

$$\widehat{\mathbf{u}}(\xi_1, y_2) = \frac{1}{\sqrt{2\pi}} \int_{\mathbb{R}} \mathbf{u}(y_1, y_2) e^{-\sqrt{-1} y_1 \xi_1} dy_1,$$

$$\widehat{\mathbf{f}}(\xi_1) = \frac{1}{\sqrt{2\pi}} \int_{\mathbb{R}} \mathbf{f}(y_1) e^{-\sqrt{-1} y_1 \xi_1} dy_1,$$

and ξ_1 is the variable of the Fourier transform. To reduce (1.124) to a first-order differential system in y_1, we introduce $\boldsymbol{\tau}(y_1, y_2)$ such that

$$\left[(\sqrt{-1}\xi_1) \mathbf{R}^T + \mathbf{T} \frac{\partial}{\partial y_2} \right] \widehat{\mathbf{u}}(\xi_1, y_2) = \sqrt{-1}\,\xi_1\, \widehat{\boldsymbol{\tau}}(\xi_1, y_2). \quad (1.125)$$

The left hand side is the Fourier transform of the traction

$$\left(\sum_{j,k,l=1}^{3} C_{ijkl} \frac{\partial u_k}{\partial x_l} n_j \right)_{i \downarrow 1,2,3}$$

in (1.24), because it can be written as

$$\left[\mathbf{R}^T \frac{\partial}{\partial y_1} + \mathbf{T} \frac{\partial}{\partial y_2} \right] \mathbf{u}(y_1, y_2).$$

As in (1.27) to (1.30), using the invertibility of $\mathbf{T}$, we see that (1.124) and (1.125) can be recast in a six-dimensional first-order differential system

$$\frac{d}{dy_2}\begin{bmatrix}\widehat{\boldsymbol{u}}\\\widehat{\boldsymbol{\tau}}\end{bmatrix} = \sqrt{-1}\,\xi_1\begin{bmatrix}-\mathbf{T}^{-1}\mathbf{R}^T & \mathbf{T}^{-1}\\-\mathbf{Q}+\mathbf{R}\mathbf{T}^{-1}\mathbf{R}^T & -\mathbf{R}\mathbf{T}^{-1}\end{bmatrix}\begin{bmatrix}\widehat{\boldsymbol{u}}\\\widehat{\boldsymbol{\tau}}\end{bmatrix}$$

$$= \sqrt{-1}\,\xi_1\,\mathbf{N}\begin{bmatrix}\widehat{\boldsymbol{u}}\\\widehat{\boldsymbol{\tau}}\end{bmatrix}. \tag{1.126}$$

The value of $\begin{bmatrix}\widehat{\boldsymbol{u}}\\\widehat{\boldsymbol{\tau}}\end{bmatrix}$ at $y_2 = 0$ is

$$\begin{bmatrix}\widehat{\boldsymbol{u}}\\\widehat{\boldsymbol{\tau}}\end{bmatrix}(\xi_1, 0) = \begin{bmatrix}\widehat{\boldsymbol{f}}(\xi_1)\\\widehat{\boldsymbol{\tau}}(\xi_1, 0)\end{bmatrix}. \tag{1.127}$$

Hence the solution to (1.126) and (1.127) is obtained as

$$\begin{bmatrix}\widehat{\boldsymbol{u}}\\\widehat{\boldsymbol{\tau}}\end{bmatrix}(\xi_1, y_2) = e^{\sqrt{-1}\,\xi_1\,\mathbf{N}\,y_2}\begin{bmatrix}\widehat{\boldsymbol{f}}(\xi_1)\\\widehat{\boldsymbol{\tau}}(\xi_1, 0)\end{bmatrix}. \tag{1.128}$$

First we consider the case where the eigenvalue problem (1.29) is simple or semisimple, and justifiy the form (1.45). Let $\begin{bmatrix}\mathbf{a}_\alpha\\\mathbf{l}_\alpha\end{bmatrix} \in \mathbb{C}^6$ ($\alpha = 1, 2, 3$) be linearly independent eigenvectors of $\mathbf{N}$ pertaining to the eigenvalues p_α ($\alpha = 1, 2, 3$, $\mathrm{Im}\,p_\alpha > 0$). We use the conventions (1.43) and (1.44). Let

$$\mathbf{P}_\alpha : \mathbb{C}^6 \longrightarrow \{\mathbf{v} \in \mathbb{C}^6 ; (\mathbf{N} - p_\alpha\mathbf{I})\mathbf{v} = 0\} \quad (1 \le \alpha \le 6) \tag{1.129}$$

be the projection operators on the eigenspaces of $\mathbf{N}$ associated with p_α, where $\mathbf{I}$ is the 6×6 identity matrix. Then

$$\sum_{\alpha=1}^{6}\mathbf{P}_\alpha = \mathbf{I}$$

and (1.129) implies that

$$\mathbf{N} = \sum_{\alpha=1}^{6}\mathbf{N}\mathbf{P}_\alpha = \sum_{\alpha=1}^{6}\left((\mathbf{N} - p_\alpha\mathbf{I})\mathbf{P}_\alpha + p_\alpha\mathbf{P}_\alpha\right) = \sum_{\alpha=1}^{6}p_\alpha\mathbf{P}_\alpha.$$

This is the spectral representation of $\mathbf{N}$.[18] Then since $\mathbf{P}_\alpha\mathbf{P}_\beta = \delta_{\alpha\beta}\mathbf{P}_\alpha$ for any α, β ($1 \le \alpha, \beta \le 6$), it follows that

$$e^{\sqrt{-1}\,\xi_1\,\mathbf{N}\,y_2} = \sum_{k=0}^{\infty}\frac{\left(\sqrt{-1}\,\xi_1\,y_2\right)^k}{k!}\mathbf{N}^k = \sum_{\alpha=1}^{6}\sum_{k=0}^{\infty}\frac{\left(\sqrt{-1}\,\xi_1\,y_2\right)^k}{k!}p_\alpha^k\,\mathbf{P}_\alpha$$

$$= \sum_{\alpha=1}^{6}e^{\sqrt{-1}\,\xi_1\,p_\alpha\,y_2}\,\mathbf{P}_\alpha.$$

---

[18]When the eigenvalue problem (1.29) is simple, the spectral representation of $\mathbf{N}$ is given in Section 4.A and 4.B of [22] explicitly in terms of eigenvectors of $\mathbf{N}$.

Hence (1.128) becomes

$$\begin{bmatrix} \widehat{\boldsymbol{u}} \\ \widehat{\boldsymbol{\tau}} \end{bmatrix} (\xi_1, y_2) = \sum_{\alpha=1}^{6} e^{\sqrt{-1}\,\xi_1\, p_\alpha\, y_2}\, \mathbf{P}_\alpha \begin{bmatrix} \widehat{\boldsymbol{f}}(\xi_1) \\ \widehat{\boldsymbol{\tau}}(\xi_1, 0) \end{bmatrix}.$$

Since Im $p_\alpha > 0$ for $\alpha = 1, 2, 3$ and Im $p_\alpha < 0$ for $\alpha = 4, 5, 6$, the boundedness of the solution as $y_2 \longrightarrow -\infty$ requires that

$$\xi_1 > 0 \quad \Longrightarrow \quad \mathbf{P}_\alpha \begin{bmatrix} \widehat{\boldsymbol{f}}(\xi_1) \\ \widehat{\boldsymbol{\tau}}(\xi_1, 0) \end{bmatrix} = \mathbf{0} \quad \text{for } \alpha = 1, 2, 3,$$

which is equivalent to

$$\begin{bmatrix} \widehat{\boldsymbol{f}}(\xi_1) \\ \widehat{\boldsymbol{\tau}}(\xi_1, 0) \end{bmatrix} \in \mathbf{P}_4\left(\mathbb{C}^6\right) \oplus \mathbf{P}_5\left(\mathbb{C}^6\right) \oplus \mathbf{P}_6\left(\mathbb{C}^6\right)$$

$$= \left\{ c_4 \begin{bmatrix} \mathbf{a}_4 \\ \mathbf{l}_4 \end{bmatrix} + c_5 \begin{bmatrix} \mathbf{a}_5 \\ \mathbf{l}_5 \end{bmatrix} + c_6 \begin{bmatrix} \mathbf{a}_6 \\ \mathbf{l}_6 \end{bmatrix} ; \ c_4, c_5, c_6 \in \mathbb{C} \right\} \quad \text{for } \xi_1 > 0,$$

$$(1.130)$$

and requires that

$$\xi_1 < 0 \quad \Longrightarrow \quad \mathbf{P}_\alpha \begin{bmatrix} \widehat{\boldsymbol{f}}(\xi_1) \\ \widehat{\boldsymbol{\tau}}(\xi_1, 0) \end{bmatrix} = \mathbf{0} \quad \text{for } \alpha = 4, 5, 6,$$

which is equivalent to

$$\begin{bmatrix} \widehat{\boldsymbol{f}}(\xi_1) \\ \widehat{\boldsymbol{\tau}}(\xi_1, 0) \end{bmatrix} \in \mathbf{P}_1\left(\mathbb{C}^6\right) \oplus \mathbf{P}_2\left(\mathbb{C}^6\right) \oplus \mathbf{P}_3\left(\mathbb{C}^6\right)$$

$$= \left\{ c_1 \begin{bmatrix} \mathbf{a}_1 \\ \mathbf{l}_1 \end{bmatrix} + c_2 \begin{bmatrix} \mathbf{a}_2 \\ \mathbf{l}_2 \end{bmatrix} + c_3 \begin{bmatrix} \mathbf{a}_3 \\ \mathbf{l}_3 \end{bmatrix} ; \ c_1, c_2, c_3 \in \mathbb{C} \right\} \quad \text{for } \xi_1 < 0;$$

$$(1.131)$$

here $\oplus$ denotes the direct sum of the vector spaces.

To establish (1.130) and (1.131), we first note that there exist scalar functions $c_\alpha(\xi_1)$ $(1 \le \alpha \le 6)$ such that

$$\widehat{\boldsymbol{f}}(\xi_1) = c_4(\xi_1)\, \mathbf{a}_4 + c_5(\xi_1)\, \mathbf{a}_5 + c_6(\xi_1)\, \mathbf{a}_6 \quad \text{for } \xi_1 > 0$$

and

$$\widehat{\boldsymbol{f}}(\xi_1) = c_1(\xi_1)\, \mathbf{a}_1 + c_2(\xi_1)\, \mathbf{a}_2 + c_3(\xi_1)\, \mathbf{a}_3 \quad \text{for } \xi_1 < 0$$

for any given boundary data $\boldsymbol{f}(y_1)$, since $\mathbf{a}_\alpha$ $(\alpha = 1, 2, 3)$ are complete in $\mathbb{C}^3$ by Theorem 1.16 and $\mathbf{a}_{\alpha+3} = \overline{\mathbf{a}_\alpha}$ $(\alpha = 1, 2, 3)$. Extending $c_\alpha(\xi_1)$ $(1 \le \alpha \le 6)$ to be functions on $\mathbb{R}$ by

$$c_\alpha(\xi_1) = 0 \ (\alpha = 1, 2, 3) \quad \text{for } \xi_1 > 0,$$

$$c_\alpha(\xi_1) = 0 \ (\alpha = 4, 5, 6) \quad \text{for } \xi_1 < 0, \qquad (1.132)$$

we can write

$$\widehat{\boldsymbol{f}}(\xi_1) = \sum_{\alpha=1}^{6} c_\alpha(\xi_1)\, \mathbf{a}_\alpha \quad (\xi_1 \neq 0).$$

Put

$$\begin{bmatrix} \widehat{\boldsymbol{u}} \\ \widehat{\boldsymbol{\tau}} \end{bmatrix}(\xi_1, y_2) = \sum_{\alpha=1}^{6} e^{\sqrt{-1}\,\xi_1\, p_\alpha\, y_2}\, \mathbf{P}_\alpha \left(\sum_{\beta=1}^{6} c_\beta(\xi_1) \begin{bmatrix} \mathbf{a}_\beta \\ \mathbf{l}_\beta \end{bmatrix} \right).$$

Then it follows that

$$\begin{bmatrix} \widehat{\boldsymbol{u}} \\ \widehat{\boldsymbol{\tau}} \end{bmatrix}(\xi_1, y_2) = \sum_{\alpha=1}^{6} e^{\sqrt{-1}\,\xi_1\, p_\alpha\, y_2}\, c_\alpha(\xi_1) \begin{bmatrix} \mathbf{a}_\alpha \\ \mathbf{l}_\alpha \end{bmatrix}.$$

Substituting the vector

$$\widehat{\boldsymbol{u}}(\xi_1, y_2) = \sum_{\alpha=1}^{6} e^{\sqrt{-1}\,\xi_1\, p_\alpha\, y_2}\, c_\alpha(\xi_1)\, \mathbf{a}_\alpha \tag{1.133}$$

which constitutes the first three components of the above into the left hand side of (1.125), we get

$$\left[(\sqrt{-1}\,\xi_1)\mathbf{R}^T + \mathbf{T}\frac{\partial}{\partial y_2} \right] \widehat{\boldsymbol{u}}(\xi_1, y_2) = \sqrt{-1}\,\xi_1 \sum_{\alpha=1}^{6} e^{\sqrt{-1}\,\xi_1\, p_\alpha\, y_2}\, c_\alpha(\xi_1) \left[\mathbf{R}^T + p_\alpha \mathbf{T} \right] \mathbf{a}_\alpha,$$

which, by (1.25), is equal to

$$\sqrt{-1}\,\xi_1 \sum_{\alpha=1}^{6} e^{\sqrt{-1}\,\xi_1\, p_\alpha\, y_2}\, c_\alpha(\xi_1)\, \mathbf{l}_\alpha.$$

Thus,

$$\widehat{\boldsymbol{\tau}}(\xi_1, 0) = \sum_{\alpha=1}^{6} c_\alpha(\xi_1)\, \mathbf{l}_\alpha.$$

This establishes (1.130) and (1.131).

Applying the inverse Fourier transform to (1.133) and using (1.121), we obtain

$$\boldsymbol{u}(y_1, y_2) = \frac{1}{\sqrt{2\pi}} \sum_{\alpha=1}^{6} \int_{\mathbb{R}} e^{\sqrt{-1}\,\xi_1\,(p_\alpha\, y_2 + y_1)}\, c_\alpha(\xi_1)\, \mathbf{a}_\alpha\, d\xi_1$$

$$= \frac{1}{\sqrt{2\pi}} \sum_{\alpha=1}^{6} \int_{\mathbb{R}} e^{\sqrt{-1}\,\xi_1\,(\mathbf{m}\cdot\mathbf{x} + p_\alpha \mathbf{n}\cdot\mathbf{x})}\, c_\alpha(\xi_1)\, \mathbf{a}_\alpha\, d\xi_1. \tag{1.134}$$

Hence $\boldsymbol{u}$ is the Fourier integral with component

$$\sum_{\alpha=1}^{6} c_\alpha(\xi_1)\, \mathbf{a}_\alpha\, e^{\sqrt{-1}\,\xi_1\,(\mathbf{m}\cdot\mathbf{x} + p_\alpha \mathbf{n}\cdot\mathbf{x})},$$

which becomes for $\xi_1 < 0$,

$$\sum_{\alpha=1}^{3} c_\alpha(\xi_1)\, \mathbf{a}_\alpha\, e^{\sqrt{-1}\,\xi_1(\mathbf{m}\cdot\mathbf{x}+p_\alpha\mathbf{n}\cdot\mathbf{x})}, \tag{1.135}$$

and becomes for $\xi_1 > 0$,

$$\sum_{\alpha=4}^{6} c_\alpha(\xi_1)\, \mathbf{a}_\alpha\, e^{\sqrt{-1}\,\xi_1(\mathbf{m}\cdot\mathbf{x}+p_\alpha\mathbf{n}\cdot\mathbf{x})}.$$

Putting $\xi_1 = -\xi$ in (1.135) gives (1.45).

Next we consider the Case D1, where the eigenvalue problem (1.29) is degenerate and there appear Jordan chains of length 2 (see Section 1.4). We justify the form (1.48). Let $\begin{bmatrix}\mathbf{a}_\alpha\\\mathbf{l}_\alpha\end{bmatrix}$ ($\alpha = 1, 2$) be linearly independent eigenvectors of $\mathbf{N}$ associated with the eigenvalues p_α ($\alpha = 1, 2$, $\mathrm{Im}\, p_\alpha > 0$), and let $\begin{bmatrix}\mathbf{a}_3\\\mathbf{l}_3\end{bmatrix}$ be a generalized eigenvector which satisfies

$$\mathbf{N}\begin{bmatrix}\mathbf{a}_3\\\mathbf{l}_3\end{bmatrix} - p_2\begin{bmatrix}\mathbf{a}_3\\\mathbf{l}_3\end{bmatrix} = \begin{bmatrix}\mathbf{a}_2\\\mathbf{l}_2\end{bmatrix}. \tag{1.136}$$

Likewise we use the conventions (1.43) for $\alpha = 1, 2$ and (1.44).

Let

$$\mathbf{P}_\alpha : \mathbb{C}^6 \longrightarrow \{\mathbf{v} \in \mathbb{C}^6\,;\, (\mathbf{N} - p_\alpha\mathbf{I})\mathbf{v} = \mathbf{0}\}, \quad \alpha = 1, 4,$$

$$\mathbf{P}_\alpha : \mathbb{C}^6 \longrightarrow \{\mathbf{v} \in \mathbb{C}^6\,;\, (\mathbf{N} - p_\alpha\mathbf{I})^2\mathbf{v} = \mathbf{0}\}, \quad \alpha = 2, 5 \tag{1.137}$$

be the projection operators on the eigenspaces of $\mathbf{N}$ ($\alpha = 1, 4$) or on the generalized eigenspaces of $\mathbf{N}$ ($\alpha = 2, 5$) associated with p_α. Then it follows that

$$\sum_{\alpha=1,2,4,5} \mathbf{P}_\alpha = \mathbf{I}$$

and

$$e^{\sqrt{-1}\,\xi_1\,\mathbf{N}\,y_2} = \sum_{\alpha=1,2,4,5} e^{\sqrt{-1}\,\xi_1\,\mathbf{N}\,y_2}\, \mathbf{P}_\alpha$$

$$= \sum_{\alpha=1,2,4,5} e^{\sqrt{-1}\,\xi_1\,p_\alpha\,y_2}\, e^{\sqrt{-1}\,\xi_1(\mathbf{N}-p_\alpha\mathbf{I})y_2}\, \mathbf{P}_\alpha$$

$$= \sum_{\alpha=1,2,4,5} e^{\sqrt{-1}\,\xi_1\,p_\alpha\,y_2}\, \sum_{k=0}^{\infty} \frac{(\sqrt{-1}\,\xi_1\,y_2)^k}{k!}\,(\mathbf{N} - p_\alpha\mathbf{I})^k\, \mathbf{P}_\alpha.$$

Then (1.137) implies that[19]

$$e^{\sqrt{-1}\,\xi_1\,\mathbf{N}\,y_2} = \sum_{\alpha=1,2,4,5} e^{\sqrt{-1}\,\xi_1\,p_\alpha\,y_2}\, \mathbf{P}_\alpha + \sum_{\alpha=2,5} e^{\sqrt{-1}\,\xi_1\,p_\alpha\,y_2}\, \sqrt{-1}\,\xi_1\,y_2(\mathbf{N} - p_\alpha\mathbf{I})\, \mathbf{P}_\alpha.$$

[19]It is also possible to obtain this formula from the spectral representation of $\mathbf{N}$ as before. But here we have adopted a more straightforward method.

Therefore (1.128) becomes

$$\begin{bmatrix} \widehat{u} \\ \widehat{\tau} \end{bmatrix} (\xi_1, y_2) = \left(\sum_{\alpha=1,2,4,5} e^{\sqrt{-1}\,\xi_1\, p_\alpha\, y_2}\, \mathbf{P}_\alpha \right.$$

$$\left. + \sum_{\alpha=2,5} e^{\sqrt{-1}\,\xi_1\, p_\alpha\, y_2}\, \sqrt{-1}\,\xi_1\, y_2\, (\mathbf{N} - p_\alpha \mathbf{I})\, \mathbf{P}_\alpha \right) \begin{bmatrix} \widehat{f}(\xi_1) \\ \widehat{\tau}(\xi_1, 0) \end{bmatrix}.$$

Since Im $p_\alpha > 0$ for $\alpha = 1, 2$ and Im $p_\alpha < 0$ for $\alpha = 4, 5$, the boundedness of the solution as $y_2 \longrightarrow -\infty$ requires that

$$\xi_1 > 0 \quad \Longrightarrow \quad \mathbf{P}_\alpha \begin{bmatrix} \widehat{f}(\xi_1) \\ \widehat{\tau}(\xi_1, 0) \end{bmatrix} = \mathbf{0} \quad \text{for } \alpha = 1, 2,$$

which is equivalent to

$$\begin{bmatrix} \widehat{f}(\xi_1) \\ \widehat{\tau}(\xi_1, 0) \end{bmatrix} \in \mathbf{P}_4(\mathbb{C}^6) \oplus \mathbf{P}_5(\mathbb{C}^6)$$

$$= \left\{ c_4 \begin{bmatrix} \mathbf{a}_4 \\ \mathbf{l}_4 \end{bmatrix} + c_5 \begin{bmatrix} \mathbf{a}_5 \\ \mathbf{l}_5 \end{bmatrix} + c_6 \begin{bmatrix} \mathbf{a}_6 \\ \mathbf{l}_6 \end{bmatrix} ; \ c_4, c_5, c_6 \in \mathbb{C} \right\} \quad \text{for } \xi_1 > 0,$$

$$(1.138)$$

and requires that

$$\xi_1 < 0 \quad \Longrightarrow \quad \mathbf{P}_\alpha \begin{bmatrix} \widehat{f}(\xi_1) \\ \widehat{\tau}(\xi_1, 0) \end{bmatrix} = \mathbf{0} \quad \text{for } \alpha = 4, 5,$$

which is equivalent to

$$\begin{bmatrix} \widehat{f}(\xi_1) \\ \widehat{\tau}(\xi_1, 0) \end{bmatrix} \in \mathbf{P}_1(\mathbb{C}^6) \oplus \mathbf{P}_2(\mathbb{C}^6)$$

$$= \left\{ c_1 \begin{bmatrix} \mathbf{a}_1 \\ \mathbf{l}_1 \end{bmatrix} + c_2 \begin{bmatrix} \mathbf{a}_2 \\ \mathbf{l}_2 \end{bmatrix} + c_3 \begin{bmatrix} \mathbf{a}_3 \\ \mathbf{l}_3 \end{bmatrix} ; \ c_1, c_2, c_3 \in \mathbb{C} \right\} \quad \text{for } \xi_1 < 0.$$

$$(1.139)$$

To establish (1.138) and (1.139), we see that there exist scalar functions $c_\alpha(\xi_1)$ ($1 \le \alpha \le 6$) such that

$$\widehat{f}(\xi_1) = c_4(\xi_1)\, \mathbf{a}_4 + c_5(\xi_1)\, \mathbf{a}_5 + c_6(\xi_1)\, \mathbf{a}_6 \quad \text{for } \xi_1 > 0$$

and

$$\widehat{f}(\xi_1) = c_1(\xi_1)\, \mathbf{a}_1 + c_2(\xi_1)\, \mathbf{a}_2 + c_3(\xi_1)\, \mathbf{a}_3 \quad \text{for } \xi_1 < 0$$

for any given boundary data $\boldsymbol{f}(y_1)$, since $\mathbf{a}_\alpha$ $(\alpha = 1, 2, 3)$ are complete in $\mathbb{C}^3$ by Theorem 1.16.[20] Then extending $c_\alpha(\xi_1)$ $(1 \leq \alpha \leq 6)$ by (1.132), we write

$$\widehat{\boldsymbol{f}}(\xi_1) = \sum_{\alpha=1}^{6} c_\alpha(\xi_1)\, \mathbf{a}_\alpha \quad (\xi_1 \neq 0).$$

Put

$$\begin{bmatrix} \widehat{\boldsymbol{u}} \\ \widehat{\boldsymbol{\tau}} \end{bmatrix}(\xi_1, y_2) = \left(\sum_{\alpha=1,2,4,5} e^{\sqrt{-1}\,\xi_1\, p_\alpha\, y_2}\, \mathbf{P}_\alpha \right.$$

$$\left. + \sum_{\alpha=2,5} e^{\sqrt{-1}\,\xi_1\, p_\alpha\, y_2}\, \sqrt{-1}\,\xi_1\, y_2\, (\mathbf{N} - p_\alpha \mathbf{I})\, \mathbf{P}_\alpha \right) \left(\sum_{\beta=1}^{6} c_\beta(\xi_1) \begin{bmatrix} \mathbf{a}_\beta \\ \mathbf{l}_\beta \end{bmatrix} \right).$$

$$(1.140)$$

Since it follows that

$$\mathbf{P}_\alpha \left(\sum_{\beta=1}^{6} c_\beta(\xi_1) \begin{bmatrix} \mathbf{a}_\beta \\ \mathbf{l}_\beta \end{bmatrix} \right) = c_\alpha(\xi_1) \begin{bmatrix} \mathbf{a}_\alpha \\ \mathbf{l}_\alpha \end{bmatrix} + c_{\alpha+1}(\xi_1) \begin{bmatrix} \mathbf{a}_{\alpha+1} \\ \mathbf{l}_{\alpha+1} \end{bmatrix} \quad (\alpha = 2, 5)$$

and from (1.136) that

$$(\mathbf{N} - p_\alpha \mathbf{I})\, \mathbf{P}_\alpha \left(\sum_{\beta=1}^{6} c_\beta(\xi_1) \begin{bmatrix} \mathbf{a}_\beta \\ \mathbf{l}_\beta \end{bmatrix} \right) = c_{\alpha+1}(\xi_1) \begin{bmatrix} \mathbf{a}_\alpha \\ \mathbf{l}_\alpha \end{bmatrix} \quad (\alpha = 2, 5),$$

formula (1.140) becomes

$$\begin{bmatrix} \widehat{\boldsymbol{u}} \\ \widehat{\boldsymbol{\tau}} \end{bmatrix}(\xi_1, y_2) = \sum_{\alpha=1,2,4,5} e^{\sqrt{-1}\,\xi_1\, p_\alpha\, y_2}\, c_\alpha(\xi_1) \begin{bmatrix} \mathbf{a}_\alpha \\ \mathbf{l}_\alpha \end{bmatrix}$$

$$+ \sum_{\alpha=2,5} e^{\sqrt{-1}\,\xi_1\, p_\alpha\, y_2}\, c_{\alpha+1}(\xi_1) \left(\begin{bmatrix} \mathbf{a}_{\alpha+1} \\ \mathbf{l}_{\alpha+1} \end{bmatrix} + \sqrt{-1}\,\xi_1\, y_2 \begin{bmatrix} \mathbf{a}_\alpha \\ \mathbf{l}_\alpha \end{bmatrix} \right).$$

Substituting the vector

$$\widehat{\boldsymbol{u}}(\xi_1, y_2) = \sum_{\alpha=1,2,4,5} e^{\sqrt{-1}\,\xi_1\, p_\alpha\, y_2}\, c_\alpha(\xi_1)\, \mathbf{a}_\alpha$$

$$+ \sum_{\alpha=2,5} e^{\sqrt{-1}\,\xi_1\, p_\alpha\, y_2}\, c_{\alpha+1}(\xi_1) \left(\mathbf{a}_{\alpha+1} + \sqrt{-1}\,\xi_1\, y_2\, \mathbf{a}_\alpha \right) \quad (1.141)$$

---

[20] Recall that Theorem 1.16 holds not only when $\begin{bmatrix} \mathbf{a}_\alpha \\ \mathbf{l}_\alpha \end{bmatrix}$ $(\alpha = 1, 2, 3)$ are linearly independent eigenvectors of $\mathbf{N}$ but also when they are linearly independent eigenvector(s) and generalized eigenvector(s) of $\mathbf{N}$.

which constitutes the first three components of the above into the left hand side of (1.125), we obtain

$$\left[(\sqrt{-1}\,\xi_1)\mathbf{R}^T + \mathbf{T}\frac{\partial}{\partial y_2}\right]\widehat{u}(\xi_1, y_2)$$

$$= \sqrt{-1}\,\xi_1 \sum_{\alpha=1,2,4,5} e^{\sqrt{-1}\,\xi_1\,p_\alpha\,y_2}\, c_\alpha(\xi_1)\left[\mathbf{R}^T + p_\alpha\mathbf{T}\right]\mathbf{a}_\alpha$$

$$+ \sqrt{-1}\,\xi_1 \sum_{\alpha=2,5} e^{\sqrt{-1}\,\xi_1\,p_\alpha\,y_2}\, c_{\alpha+1}(\xi_1)\left(\left[\mathbf{R}^T + p_\alpha\mathbf{T}\right](\mathbf{a}_{\alpha+1} + \sqrt{-1}\,\xi_1\,y_2\,\mathbf{a}_\alpha) + \mathbf{T}\,\mathbf{a}_\alpha\right),$$

which, by (1.25) and (1.50), is equal to

$$\sqrt{-1}\,\xi_1\left(\sum_{\alpha=1,2,4,5} e^{\sqrt{-1}\,\xi_1\,p_\alpha\,y_2}\, c_\alpha(\xi_1)\,\mathbf{l}_\alpha + \sum_{\alpha=2,5} e^{\sqrt{-1}\,\xi_1\,p_\alpha\,y_2}\, c_{\alpha+1}(\xi_1)\,\mathbf{l}_{\alpha+1}\right)$$

$$- \xi_1^2 \sum_{\alpha=2,5} e^{\sqrt{-1}\,\xi_1\,p_\alpha\,y_2}\, c_{\alpha+1}(\xi_1)\,y_2\,\mathbf{l}_\alpha.$$

Thus,

$$\widehat{\tau}(\xi_1, 0) = \sum_{\alpha=1}^{6} c_\alpha(\xi_1)\mathbf{l}_\alpha.$$

This establishes (1.138) and (1.139).

Applying the inverse Fourier transform to (1.141), we obtain

$$u(y_1, y_2) = \frac{1}{\sqrt{2\pi}} \sum_{\alpha=1,2,4,5} \int_{\mathbb{R}} e^{\sqrt{-1}\,\xi_1(p_\alpha\,y_2 + y_1)}\, c_\alpha(\xi_1)\,\mathbf{a}_\alpha\,d\xi_1$$

$$+ \frac{1}{\sqrt{2\pi}} \sum_{\alpha=2,5} \int_{\mathbb{R}} e^{\sqrt{-1}\,\xi_1(p_\alpha\,y_2 + y_1)}\, c_{\alpha+1}(\xi_1)\left(\mathbf{a}_{\alpha+1} + \sqrt{-1}\,\xi_1\,y_2\,\mathbf{a}_\alpha\right)d\xi_1$$

$$= \frac{1}{\sqrt{2\pi}} \sum_{\alpha=1,2,4,5} \int_{\mathbb{R}} e^{\sqrt{-1}\,\xi_1(\mathbf{m}\cdot\mathbf{x} + p_\alpha\mathbf{n}\cdot\mathbf{x})}\, c_\alpha(\xi_1)\,\mathbf{a}_\alpha\,d\xi_1$$

$$+ \frac{1}{\sqrt{2\pi}} \sum_{\alpha=2,5} \int_{\mathbb{R}} e^{\sqrt{-1}\,\xi_1(\mathbf{m}\cdot\mathbf{x} + p_\alpha\mathbf{n}\cdot\mathbf{x})}\, c_{\alpha+1}(\xi_1)\left(\mathbf{a}_{\alpha+1} + \sqrt{-1}\,\xi_1\,(\mathbf{n}\cdot\mathbf{x})\,\mathbf{a}_\alpha\right)d\xi_1.$$

$$(1.142)$$

Hence u is the Fourier integral with component

$$\sum_{\alpha=1,2,4,5} c_\alpha(\xi_1)\,\mathbf{a}_\alpha\, e^{\sqrt{-1}\,\xi_1(\mathbf{m}\cdot\mathbf{x} + p_\alpha\mathbf{n}\cdot\mathbf{x})}$$

$$+ \sum_{\alpha=2,5} c_{\alpha+1}(\xi_1)\left(\mathbf{a}_{\alpha+1} + \sqrt{-1}\,\xi_1\,(\mathbf{n}\cdot\mathbf{x})\,\mathbf{a}_\alpha\right) e^{\sqrt{-1}\,\xi_1(\mathbf{m}\cdot\mathbf{x} + p_\alpha\mathbf{n}\cdot\mathbf{x})},$$

which becomes for $\xi_1 < 0$,

$$\sum_{\alpha=1,2} c_\alpha(\xi_1) \, \mathbf{a}_\alpha \, e^{\sqrt{-1}\,\xi_1(\mathbf{m}\cdot\mathbf{x}+p_\alpha\mathbf{n}\cdot\mathbf{x})}$$

$$+ \; c_3(\xi_1) \left(\mathbf{a}_3 + \sqrt{-1}\,\xi_1 \,(\mathbf{n}\cdot\mathbf{x}) \,\mathbf{a}_2 \right) e^{\sqrt{-1}\,\xi_1(\mathbf{m}\cdot\mathbf{x}+p_2\mathbf{n}\cdot\mathbf{x})}, \qquad (1.143)$$

and becomes for $\xi_1 > 0$,

$$\sum_{\alpha=4,5} c_\alpha(\xi_1) \, \mathbf{a}_\alpha \, e^{\sqrt{-1}\,\xi_1(\mathbf{m}\cdot\mathbf{x}+p_\alpha\mathbf{n}\cdot\mathbf{x})}$$

$$+ \; c_6(\xi_1) \left(\mathbf{a}_6 + \sqrt{-1}\,\xi_1 \,(\mathbf{n}\cdot\mathbf{x}) \,\mathbf{a}_5 \right) e^{\sqrt{-1}\,\xi_1(\mathbf{m}\cdot\mathbf{x}+p_5\mathbf{n}\cdot\mathbf{x})}.$$

Putting $\xi_1 = -\xi$ in (1.143) leads to (1.48).

In the Case D2, where the eigenvalue problem (1.29) is degenerate and there appear Jordan chains of length 3, justification of the form (1.55) can be achieved by a similar method using (1.54). The details are left as Exercise 1-13.

It can be proved that the solutions $\boldsymbol{u}(y_1, y_2)$ in (1.134) and (1.142) are real when the boundary data $\boldsymbol{f}(y_1)$ in (1.123) is a real-valued function (Exercise 1-14).

1.10 Comments and References

In Section 1.1 we have summarized briefly linearized elasticity which is needed in the subsequent sections. For more details we refer to [26, 33].

We have begun our presentation of the Stroh formalism in Section 1.2. The Stroh formalism goes back to Stroh's original works [65, 66], where he studied dislocations, cracks and elastic waves in a steady state of motion in an anisotropic elastic body. Since then, many researchers extended his study to establish a mathematical theory, which is now called the Stroh formalism. The Stroh formalism reveals simple structures hidden in the equations of anisotropic elasticity and provides a systematic approach to these equations. In this formalism, isotropic elasticity is one example to which this formalism can be applied, although isotropic elasticity had been studied for a long time by using the theory proper only to it. The Stroh formalism is now indispensable to the studies of many areas in anisotropic elasticity, which include elastostatics, elastodynamics, Rayleigh waves, composite materials, cracks, inclusions and inverse problems.

In Sections 1.2 and 1.4 we have introduced the Stroh formalism, where we put emphasis on the forms of solutions (i.e., displacements) to the equations of anisotropic elasticity in a half-space and on the forms of the corresponding tractions. We have called these forms of solutions "general forms", the nomenclature of which is justified in Section 1.9 by the theory of Fourier analysis. These forms will be used in Section 2.3 when we construct approximate solutions pertaining to oscillating Dirichlet data in inverse boundary value problems.

In Sections 1.3 and 1.5 we have proved, by an intuitive method, rotational invariance and rotational dependence of the Stroh eigenvectors and generalized eigenvectors in the reference plane. Classically, this rotational dependency was proved through the fact that the 6×6 matrix $\mathbf{N}(\phi)$ in (1.33) satisfies the matrix Riccati equation

$$\frac{d}{d\phi}\mathbf{N}(\phi) = -\mathbf{I} - \mathbf{N}(\phi)^2, \qquad (1.144)$$

where $\mathbf{I}$ is the 6×6 identity matrix [9, 22, 41]. Equation (1.144) can be proved by simple but long computations where we differentiate each 3×3 block-component of $\mathbf{N}(\phi)$ with respect to ϕ and use the relations (1.35). Comparing the corresponding block-components of $\frac{d}{d\phi}\mathbf{N}(\phi)$ and $-\mathbf{I} - \mathbf{N}(\phi)^2$, we arrive at (1.144). The details are left as Exercise 1-5. We have also suggested a proof of Theorem 1.5 and of Theorem 1.10 in Exercise 1-6 and Exercise 1-7, respectively, which make use of (1.144).

It may be difficult to notice without any suggestion, that $\mathbf{N}(\phi)$ satisfies (1.144). We note that in the case where the eigenvalue problem (1.29) is simple or semisimple Theorem 1.5 suggests (1.144). In fact, formula (1.42) implies that

$$\mathbf{N}(\phi)^2 \begin{bmatrix} \mathbf{a} \\ \mathbf{l} \end{bmatrix} = p(\phi)^2 \begin{bmatrix} \mathbf{a} \\ \mathbf{l} \end{bmatrix}$$

and

$$\frac{d}{d\phi}\mathbf{N}(\phi) \begin{bmatrix} \mathbf{a} \\ \mathbf{l} \end{bmatrix} = p'(\phi) \begin{bmatrix} \mathbf{a} \\ \mathbf{l} \end{bmatrix}$$

for all ϕ, where $\begin{bmatrix} \mathbf{a} \\ \mathbf{l} \end{bmatrix}$ is an eigenvector of $\mathbf{N}(0)$. Since the eigenvalue $p(\phi)$ of $\mathbf{N}(\phi)$ satisfies the Riccati equation (1.34), it follows that

$$\frac{d}{d\phi}\mathbf{N}(\phi) \begin{bmatrix} \mathbf{a} \\ \mathbf{l} \end{bmatrix} = \left(-\mathbf{I} - \mathbf{N}(\phi)^2\right) \begin{bmatrix} \mathbf{a} \\ \mathbf{l} \end{bmatrix}.$$

For $\begin{bmatrix} \mathbf{a} \\ \mathbf{l} \end{bmatrix}$, we can take six linearly independent eigenvectors of $\mathbf{N}(0)$, which proves (1.144).

The Barnett-Lothe integral formalism was originally developed by Barnett and Lothe on the basis of the Stroh formalism ([9, 41]; see also [22]). Theorem 1.15 in Section 1.6, Theorems 1.16, 1.18, 1.21 and Lemma 1.22 in Section 1.7 are the main results in this integral formalism. Among them Theorem 1.15 is the most fundamental. In Subsection 2.2.2, we shall give a rigorous proof of a higher-order dimensional version of this theorem, which can be applied to any degenerate case of Stroh's eigenvalue problem (cf. [2, 52, 54]). We can see that the results in the integral formalism do not depend on whether the corresponding Stroh's eigenvalue problem is simple, semisimple or degenerate. Thus, the results in the integral formalism are powerful and useful in the studies on the fundamental solutions in Chapter 2 and on Rayleigh waves in Chapter 3.

In Section 1.8, we have computed explicit formulas for the surface impedance tensors on the basis of definition (1.74). Lemma 1.25 and Theorem 1.26 are the results in [69]. To our knowledge, in three-dimensional elasticity, the exact and complete formulas for the surface impedance tensors written in terms of elasticity tensors have been obtained only for isotropic and for transversely isotropic media. The difficulty is caused mainly by the fact that the characteristic equations (1.22) corresponding to the other elastic media can not be factorized. The explicit formulas for the surface impedance tensors given in this section will be used when we derive the explicit forms of the fundamental solutions and prove the identifiability of the elasticity tensor from the Dirichlet to Neumann map in the inverse boundary value problems in Chapter 2.

Note that the integral representation (1.76) of the surface impedance tensors is useful for perturbation arguments. On the basis of this representation, the surface impedance tensor is computed for a weakly anisotropic elastic medium by considering the deviation of its elasticity tensor from a comparative isotropic state as a perturbation, and an explicit perturbation formula for the surface impedance tensor is obtained in [72], which is correct to first order in the perturbative part of the elasticity tensor.

As we have seen in Section 1.8, for isotropic elasticity, Stroh's eigenvalue problem (1.29) is degenerate and it always has a Jordan chain of length of exactly two (Case D1). For transversely isotropic materials, the length of the Jordan chain is at most two. Hence there arises the question whether there exists an anisotropic elastic material for which Stroh's eigenvalue problem has a Jordan chain of length three (i.e, an elastic material which belongs to Case D2, or which is more degenerate than isotropic elasticity). Ting [78] proved that such materials exist.

Section 1.9 is devoted to justification of the forms of basic solutions (displacements) introduced in Sections 1.2 and 1.4. This justification, though straightforward, does not seem to have appeared explicitly in the literature before.

1.11 Exercises

1-1 Prove that the strong convexity condition (1.7) is equivalent to the assertion that the 6×6 matrix (1.12) in the Voigt notation is positive definite.

1-2 Find the strong convexity condition for the isotropic elasticity tensor (1.11). (Answer: (1.83))

1-3 Find the strong convexity condition for the elasticity tensor of transversely isotropic materials (1.13). (Answer: (1.101))

1-4 Prove Lemma 1.7.

1-5 Prove (1.144) by following the suggestions below it.

1-6 Use (1.144) to prove Theorem 1.5.
(Suggestion: Put

$$\mathbf{k}(\phi) = \mathbf{N}(\phi) \begin{bmatrix} \mathbf{a} \\ \mathbf{l} \end{bmatrix} - p(\phi) \begin{bmatrix} \mathbf{a} \\ \mathbf{l} \end{bmatrix}.$$

Then $\mathbf{k}(0) = \mathbf{0}$ from the assumption. Derive the ordinary differential equation for $\mathbf{k}(\phi)$ and apply the uniqueness of the solution to it as in the proof of Theorem 1.3 to show that $\mathbf{k}(\phi) = \mathbf{0}$ for all ϕ.)

1-7 Prove Theorem 1.10 in the following way:
Put

$$\mathbf{f}(\phi) = \mathbf{N}(\phi) \begin{bmatrix} \mathbf{a}_2(\phi) \\ \mathbf{l}_2(\phi) \end{bmatrix} - p_1(\phi) \begin{bmatrix} \mathbf{a}_2(\phi) \\ \mathbf{l}_2(\phi) \end{bmatrix} - \begin{bmatrix} \mathbf{a}_1 \\ \mathbf{l}_1 \end{bmatrix}$$

and

$$\chi(\phi) = \mathbf{N}(\phi) \begin{bmatrix} \mathbf{a}_3(\phi) \\ \mathbf{l}_3(\phi) \end{bmatrix} - p_1(\phi) \begin{bmatrix} \mathbf{a}_3(\phi) \\ \mathbf{l}_3(\phi) \end{bmatrix} - \begin{bmatrix} \mathbf{a}_2(\phi) \\ \mathbf{l}_2(\phi) \end{bmatrix}.$$

Then $\mathbf{f}(0) = \chi(0) = \mathbf{0}$ from the assumption. Show that

$$\mathbf{f}(\phi) = \mathbf{0}, \qquad \chi(\phi) = \mathbf{0}$$

for all ϕ.

(Suggestion: Derive the ordinary differential equations for $\mathbf{f}(\phi)$ and $\chi(\phi)$ with the help of (1.144), and apply the uniqueness of the solutions to them.)

1-8 Prove Corollary 1.19 from Theorems 1.5, 1.9 and 1.10.

1-9 Establish Remark 1.23.

1-10 Check (1.93) and complete the calculations of (1.94).

1-11 Suppose that $\mathbf{C} = \left(C_{ijkl}\right)_{i,j,k,l=1,2,3}$ does not observe any symmetry condition (i.e., it has neither the major symmetry (1.6) nor the minor symmetries (1.4)). We assume, in place of the strong convexity condition (1.7), the following ellipticity condition:

$$\det\left(\sum_{j,l=1}^{3} C_{ijkl}\,\xi_j\xi_l\right)_{i\downarrow k\to 1,2,3} \neq 0$$

for any non-zero vector $\boldsymbol{\xi} = (\xi_1, \xi_2, \xi_3) \in \mathbb{R}^3$.

In addition to the 3×3 matrices $\mathbf{Q}, \mathbf{R}, \mathbf{T}$ in (1.20), we introduce a 3×3 matrix

$$\mathbf{S} = \left(\sum_{j,l=1}^{3} C_{ijkl}\, n_j m_l\right)_{i\downarrow k\to 1,2,3}.$$

We define the 'traction' on $\mathbf{n} \cdot \mathbf{x} = 0$, i.e., the Neumann data, by (1.24) and define a vector $\mathbf{l}$ by (1.25) with $\mathbf{R}^T$ replaced by $\mathbf{S}$. Then prove Theorems 1.2, 1.3 and Theorem 1.5 with the matrix $\mathbf{R}^T$ replaced by the matrix $\mathbf{S}$.

1-12 (Two-dimensional homogeneous elasticity) Consider the plane-strain deformations of orthorhombic materials (1.15) in the (x_1, x_2)-plane. The governing equations become

$$\sum_{j,k,l=1}^{2} C_{ijkl}\frac{\partial^2 u_k}{\partial x_j \partial x_l} = 0, \qquad i = 1, 2$$

with $C_{1112} = C_{2212} = 0$. Let $\mathbf{m} = (m_1, m_2)$, $\mathbf{n} = (n_1, n_2)$ be orthogonal unit vectors in $\mathbb{R}^2$. Then, by changing the ranges where the indices i, j, k, l run from $\{1, 2, 3\}$ to $\{1, 2\}$, the arguments in Section 1.2 to Section 1.7 apply in a parallel way. We note that the dimension of the matrices $\mathbf{Q}, \mathbf{R}, \mathbf{T}, \mathbf{S}_1, \mathbf{S}_2, \mathbf{S}_3$ becomes two and the dimension of the matrices $\mathbf{N}, \mathbf{S}$ and the eigenvalue problem (1.29) becomes four. Hence the length of the Jordan chain is at most two. For simplicity, we put

$$C_{1111} = A, \quad C_{1122} = F, \quad C_{2222} = C, \quad C_{1212} = L. \tag{1.145}$$

Investigate the degeneracy of the eigenvalue problem (1.29) and compute the surface impedance tensor $\mathbf{Z}$, which is a 2×2 matrix.

Answer:

$$\mathbf{Z} = \begin{bmatrix} Z_{11} & Z_{12} \\ Z_{21} & Z_{22} \end{bmatrix}$$

 Springer

with

$$Z_{11} = \frac{\sqrt{A}\,LG}{\sqrt{AC}+L}, \qquad Z_{22} = \frac{\sqrt{C}\,LG}{\sqrt{AC}+L},$$

$$Z_{12} = -Z_{21} = \mp\frac{\sqrt{-1}(\sqrt{AC}-F)L}{\sqrt{AC}+L},$$

where

$$G = \sqrt{\frac{AC - F^2 + 2L(\sqrt{AC}-F)}{L}}$$

and the $\mp$ sign corresponds to that of $\tilde{\mathbf{m}} \times \tilde{\mathbf{n}} = (0,0,\pm 1)$ with $\tilde{\mathbf{m}} = (m_1, m_2, 0)$, $\tilde{\mathbf{n}} = (n_1, n_2, 0)$ (see [53]).

1-13 Justify the form of the solution (1.55) from (1.128) in the Case D2, where the eigenvalue problem (1.29) is degenerate and there appear Jordan chains of length 3.

1-14 Following the suggestions below, prove that the solutions $\mathbf{u}(y_1, y_2)$ in (1.134) and (1.142) are real when the boundary data $\mathbf{f}(y_1)$ in (1.123) is a real-valued function.

(a) Show that $\overline{\widehat{\mathbf{f}}(\xi_1)} = \widehat{\mathbf{f}}(-\xi_1)$ ($\xi_1 \in \mathbb{R}$).
(b) Show that $\overline{c_{\alpha+3}(\xi_1)} = c_\alpha(-\xi_1)$ ($\alpha = 1, 2, 3$) for $\xi_1 > 0$.
(c) In view of (1.43) and (1.44), recast formulas (1.134) and (1.142) using only quantities with the index α running from 1 to 3.

2 Applications in Static Elasticity

2.1 Fundamental Solutions

2.1.1 Fundamental Solution in the Stroh Formalism

Let $\mathbf{x} \in \mathbb{R}^3$ and let $G_{km} = G_{km}(\mathbf{x})$ be a solution to

$$\sum_{j,k,l=1}^{3} C_{ijkl} \frac{\partial^2}{\partial x_j \partial x_l} G_{km} + \delta_{im} \delta(\mathbf{x}) = 0 \quad \text{in } \mathbb{R}^3, \qquad i, m = 1, 2, 3, \qquad (2.1)$$

where (x_1, x_2, x_3), δ_{im}, and $\delta(\mathbf{x})$ are the Cartesian coordinates of $\mathbf{x}$, Kronecker's delta symbol, and the Dirac delta function, respectively, and $\mathbf{C} = \left(C_{ijkl}\right)_{i,j,k,l=1,2,3}$ is a homogeneous elasticity tensor. Physically, the solution G_{km} describes the displacement at the point $\mathbf{x}$ in the x_k direction due to a unit point force at the origin in the x_m direction. We call $\mathbf{G} = \mathbf{G}(\mathbf{x}) = (G_{km})_{k \downarrow m \to 1,2,3}$ the fundamental solution to

$$\sum_{j,k,l=1}^{3} C_{ijkl} \frac{\partial^2}{\partial x_j \partial x_l} u_k(\mathbf{x}) + f_i(\mathbf{x}) = 0 \quad \text{in } \mathbb{R}^3, \qquad i = 1, 2, 3;$$

it generates the solution

$$\mathbf{u}(\mathbf{x}) = (u_1, u_2, u_3) = \int_{\mathbb{R}^3} \mathbf{G}(\mathbf{x} - \mathbf{y}) \mathbf{f}(\mathbf{y}) \, d\mathbf{y},$$

where $\mathbf{f}(\mathbf{x}) = (f_1, f_2, f_3)^T$.

Let $\mathbf{x} \neq \mathbf{0}$. Take any two orthogonal unit vectors $\mathbf{e}_1$ and $\mathbf{e}_2$ in $\mathbb{R}^3$ such that

$$\mathbf{e}_1 \times \mathbf{e}_2 = \frac{\mathbf{x}}{|\mathbf{x}|},$$

where $\mathbf{e}_1 \times \mathbf{e}_2$ denotes the vector product.

Now we will use the 3×3 matrix $\mathbf{T}(\phi)$ introduced in (1.32):

$$\mathbf{T}(\phi) = \left(\sum_{j,l=1}^{3} C_{ijkl} n_j n_l \right)_{i \downarrow k \to 1,2,3},$$

where

$$\mathbf{n} = \mathbf{n}(\phi) = (n_1, n_2, n_3) = -\mathbf{e}_1 \sin \phi + \mathbf{e}_2 \cos \phi.$$

The following thoerem is well known (cf. [6, 39, 68]).

Theorem 2.1

$$\mathbf{G}(\mathbf{x}) = \frac{1}{8\pi^2 |\mathbf{x}|} \int_{-\pi}^{\pi} \mathbf{T}(\phi)^{-1} d\phi \qquad (2.2)$$

is a fundamental solution.

From (2.1), $\mathbf{G}(x)$ is given as the inverse Fourier transform of the matrix

$$\left[\left(\sum_{j,l=1}^{3} C_{ijkl}\,\xi_j \xi_l\right)_{i\downarrow k\to 1,2,3}\right]^{-1},$$

where ξ_i ($i = 1, 2, 3$) are the variables of the Fourier transform. It can be proved that this inverse Fourier transform can be reduced to a single integral over the unit circle on the plane perpendicular to $\mathbf{x}$. A rigorous proof is given in [54].

Let us write $\mathbf{G}(\mathbf{x})$ in terms of the surface impedance tensor $\mathbf{Z}$. By using the notation in (1.68), formula (2.2) can be written as

$$\mathbf{G}(\mathbf{x}) = \frac{1}{4\pi \, |\mathbf{x}|}\, \mathbf{S}_2. \tag{2.3}$$

On the other hand, from the integral representation (1.76) of $\mathbf{Z}$ it follows that

$$\mathbf{S}_2 = \left(\mathrm{Re}\,\mathbf{Z}\right)^{-1},$$

where $\mathbf{Z}$ is the surface impedance tensor defined by (1.74) (or equivalently by (1.76)) with the orthogonal unit vectors

$$\mathbf{m} = \mathbf{m}(\phi) = (m_1, m_2, m_3) = \mathbf{e}_1 \cos\phi + \mathbf{e}_2 \sin\phi,$$

$$\mathbf{n} = \mathbf{n}(\phi) = (n_1, n_2, n_3) = -\mathbf{e}_1 \sin\phi + \mathbf{e}_2 \cos\phi,$$

and $\mathrm{Re}\,\mathbf{Z}$ denotes the real part of $\mathbf{Z}$. Since it follows that

$$\mathbf{e}_1 \times \mathbf{e}_2 = \mathbf{m} \times \mathbf{n},$$

from Corollary 1.19 we obtain

Theorem 2.2

$$\mathbf{G}(\mathbf{x}) = \frac{1}{4\pi \, |\mathbf{x}|}\, \left(\mathrm{Re}\,\mathbf{Z}\right)^{-1} \tag{2.4}$$

is a fundamental solution, where $\mathbf{Z}$ *is the surface impedance tensor in (1.74) with* $\mathbf{m}$ *and* $\mathbf{n}$ *satisfying*

$$\mathbf{m} \times \mathbf{n} = \frac{\mathbf{x}}{|\mathbf{x}|}.$$

2.1.2 Formulas for Fundamental Solutions: Examples

In Section 1.8, we have obtained explicit formulas for the surface impedance tensors of isotropic and transversely isotropic materials. In this subsection, applying Theorem 2.2 to these formulas, we derive the formulas for the fundamental solutions pertaining to isotropic and transversely isotropic materials.

For $\mathbf{x} \neq 0$, we write

$$\frac{\mathbf{x}}{|\mathbf{x}|} = (\widehat{x}_1, \widehat{x}_2, \widehat{x}_3).$$

For isotropic materials, from (1.95) and (2.4) we immediately obtain

 Springer

Corollary 2.3 (*Fundamental solution for isotropic materials*)

$$\mathbf{G}^{\mathrm{Iso}}(\mathbf{x}) = \left(\mathbf{G}^{\mathrm{Iso}}\right)^T(\mathbf{x}) = \left(G_{ij}\right)_{i\downarrow j \to 1,2,3},$$

$$G_{ij} = \frac{1}{8\pi\,\mu(\lambda + 2\mu)|\mathbf{x}|}\left((\lambda + 3\mu)\delta_{ij} + (\lambda + \mu)\widehat{x}_i\widehat{x}_j\right).$$

For transversely isotropic materials, using the notations (1.100), we obtain

Corollary 2.4 (*Fundamental solution for transversely isotropic materials*)

$$\mathbf{G}^{\mathrm{Trans}}(\mathbf{x}) = \left(\mathbf{G}^{\mathrm{Trans}}\right)^T(\mathbf{x}) = \frac{1}{4\pi\,|\mathbf{x}|}\left(S_{ij}\right)_{i\downarrow j \to 1,2,3},$$

$$S_{11} = S_{11}^0\frac{\widehat{x}_1^2}{\widehat{x}_1^2 + \widehat{x}_2^2} + S_{22}^0\frac{\widehat{x}_2^2}{\widehat{x}_1^2 + \widehat{x}_2^2}, \qquad S_{22} = S_{11}^0\frac{\widehat{x}_2^2}{\widehat{x}_1^2 + \widehat{x}_2^2} + S_{22}^0\frac{\widehat{x}_1^2}{\widehat{x}_1^2 + \widehat{x}_2^2},$$

$$S_{12} = \left(S_{11}^0 - S_{22}^0\right)\frac{\widehat{x}_1\widehat{x}_2}{\widehat{x}_1^2 + \widehat{x}_2^2},$$

$$S_{13} = \frac{F + L}{ALGH}\widehat{x}_1\widehat{x}_3, \qquad S_{23} = \frac{F + L}{ALGH}\widehat{x}_2\widehat{x}_3.$$

$$S_{33} = \frac{1}{ALGH}\left(A\left(\widehat{x}_3^2 + H\right) + L\left(\widehat{x}_1^2 + \widehat{x}_2^2\right)\right), \tag{2.5}$$

where

$$S_{11}^0 = \frac{AL\left(GHK - \widehat{x}_3^4 - H\widehat{x}_3^2\right) - \left(AC - (F + L)^2\right)\left(\widehat{x}_1^2 + \widehat{x}_2^2\right)\widehat{x}_3^2}{AL^2GH\left(\widehat{x}_1^2 + \widehat{x}_2^2\right)},$$

$$S_{22}^0 = \frac{\left(A\widehat{x}_3^2 + L\left(\widehat{x}_1^2 + \widehat{x}_2^2\right) + AH\right)K - AG\widehat{x}_3^2}{AGKL\left(\widehat{x}_1^2 + \widehat{x}_2^2\right)}$$

and K, G and H are given by (1.104) *with* $\ell_i = \widehat{x}_i$ (*i* = 1, 2, 3), *that is,*

$$K = \sqrt{\widehat{x}_3^2 + \frac{2L}{A - N}\left(\widehat{x}_1^2 + \widehat{x}_2^2\right)},$$

$$G = \sqrt{\frac{2AL\widehat{x}_3^2 + (AC - F^2 - 2FL)\left(\widehat{x}_1^2 + \widehat{x}_2^2\right) + 2\sqrt{AL}\sqrt{\Delta}}{AL}},$$

$$H = \sqrt{\frac{\Delta}{AL}}, \qquad \Delta = AL\widehat{x}_3^4 + (AC - F^2 - 2FL)\left(\widehat{x}_1^2 + \widehat{x}_2^2\right)\widehat{x}_3^2 + CL\left(\widehat{x}_1^2 + \widehat{x}_2^2\right)^2.$$

Note In the case $\widehat{x}_3 = \pm 1$, i.e., the case $\widehat{x}_1^2 + \widehat{x}_2^2 = 0$, S_{11}^0 and S_{22}^0 become indeterminate forms. In this case, as in the note of Theorem 1.26, $\mathbf{G}^{\text{Trans}}(\mathbf{x})$ is obtained by taking the limit $\widehat{x}_1^2 + \widehat{x}_2^2 \longrightarrow 0$ in the formula above. In fact, as $\widehat{x}_1^2 + \widehat{x}_2^2 \longrightarrow 0$ we get

$$S_{11}^0 \longrightarrow \frac{3A - N}{2A(A - N)}, \qquad S_{22}^0 \longrightarrow \frac{3A - N}{2A(A - N)}.$$

Then the components of $(S_{ij})_{i\downarrow j \to 1,2,3}$ in (2.5) in the case $\widehat{x}_1^2 + \widehat{x}_2^2 = 0$ become

$$S_{11} = S_{22} = \frac{3A - N}{2A(A - N)}, \qquad S_{33} = \frac{1}{L}, \qquad S_{12} = S_{13} = S_{23} = 0.$$

Recall that in (1.100), the 3-axis is the axis of rotational symmetry. Hence $\widehat{x}_3 = \pm 1$ implies that $\mathbf{x}$ lies on the axis of rotational symmetry.

Outline of Proof of Corollary 2.4 Using spherical coordinates, we write

$$\frac{\mathbf{x}}{|\mathbf{x}|} = (\widehat{x}_1, \widehat{x}_2, \widehat{x}_3) = (\sin\alpha\,\cos\beta, \sin\alpha\,\sin\beta, \cos\alpha)$$

$$(0 \le \alpha \le \pi, 0 \le \beta < 2\pi). \tag{2.6}$$

Instead of applying (2.4) to (1.103) directly, as in Subsection 1.8.2 we make use of the transformation formula for $\mathbf{G}^{\text{Trans}}(\mathbf{x})$

$$\mathbf{G}^{\text{Trans}}(\mathbf{x}) = \Omega_3\,\mathbf{G}^{\text{Trans}}(\mathbf{x}_0)\,\Omega_3^T, \tag{2.7}$$

where

$$|\mathbf{x}_0| = |\mathbf{x}|, \qquad \frac{\mathbf{x}_0}{|\mathbf{x}_0|} = (\sin\alpha, 0, \cos\alpha),$$

and Ω_3 is the transformation matrix (1.97) pertaining to the rotation of the material around the 3-axis by the angle β. Theorem 2.2 implies that

$$\mathbf{G}^{\text{Trans}}(\mathbf{x}_0) = \frac{1}{4\pi\,|\mathbf{x}|}\,(\operatorname{Re}\mathbf{Z}_0)^{-1},$$

where $\mathbf{Z}_0$ is the surface impedance tensor for $\mathbf{m} \times \mathbf{n} = (\sin\alpha, 0, \cos\alpha)$ and its explicit form is already given by (1.118). Then by simple but long conputations we get

$$\mathbf{G}^{\text{Trans}}(\mathbf{x}_0) = (\mathbf{G}^{\text{Trans}})^T(\mathbf{x}_0) = \frac{1}{4\pi\,|\mathbf{x}|}\,(S_{ij}^0)_{i\downarrow j \to 1,2,3},$$

$$S_{11}^0 = \frac{AL(GHK - \cos^4\alpha - H\cos^2\alpha) - (AC - (F + L)^2)\cos^2\alpha\,\sin^2\alpha}{AL^2GH\,\sin^2\alpha},$$

$$S_{22}^0 = \frac{(A\cos^2\alpha + L\sin^2\alpha + AH)\,K - AG\cos^2\alpha}{AGKL\,\sin^2\alpha},$$

$$S_{33}^0 = \frac{1}{ALGH}\,(A(\cos^2\alpha + H) + L\sin^2\alpha),$$

$$S_{12}^0 = 0, \qquad S_{13}^0 = \frac{F + L}{ALGH}\,\cos\alpha\,\sin\alpha, \qquad S_{23}^0 = 0. \tag{2.8}$$

Thus, from (2.6) and (2.7) we obtain the result. □

2.2 Piezoelectricity

2.2.1 Basic Theory

Piezoelectric materials have been used in many engineering devices because of their intrinsic direct and converse piezoelectric effects that take place between electric fields and mechanical deformations. In piezoelectricity, the mechanical stress tensor $\sigma = (\sigma_{ij})_{i,j=1,2,3}$ and the electric displacement $D = (D_1, D_2, D_3)$ are related to the mechanical displacement $u = (u_1, u_2, u_3)$ and the electric potential ϕ by the following constitutive equations:

$$\sigma_{ij} = \sum_{k,l=1}^{3} C_{ijkl} \frac{\partial u_k}{\partial x_l} + \sum_{l=1}^{3} e_{ijl} \frac{\partial \phi}{\partial x_l}, \qquad i,j = 1,2,3,$$

$$D_j = \sum_{k,l=1}^{3} e_{klj} \frac{\partial u_k}{\partial x_l} - \sum_{l=1}^{3} \epsilon_{jl} \frac{\partial \phi}{\partial x_l}, \qquad j = 1,2,3. \tag{2.9}$$

Here $\mathbf{C} = (C_{ijkl})_{i,j,k,l=1,2,3}$ is the elasticity tensor, $e = (e_{ijl})_{i,jl=1,2,3}$ is the piezoelectric tensor, and $\epsilon = (\epsilon_{jl})_{i,l=1,2,3}$ is the dielectric tensor. Hence the elastic and electric fields are coupled through the piezoelectric tensor. These tensors $\mathbf{C}$, e, and ϵ satisfy the following symmetry conditions:

$$C_{ijkl} = C_{jikl} = C_{klij}, \qquad e_{ijl} = e_{jil}, \qquad \epsilon_{jl} = \epsilon_{lj}, \qquad i,j,k,l = 1,2,3. \tag{2.10}$$

We assume that the internal energy function

$$U = \sum_{i,j,k,l=1}^{3} \frac{1}{2} C_{ijkl}\, \varepsilon_{ij}\, \varepsilon_{kl} + \sum_{j,l=1}^{3} \frac{1}{2} \epsilon_{jl}\, E_j\, E_l$$

be positive for any non-zero mechanical strain $(\varepsilon_{ij})_{i,j=1,2,3}$ and for any non-zero electric field (E_1, E_2, E_3). This leads to the following strong convexity conditions:

$$\sum_{i,j,k,l=1}^{3} C_{ijkl}\, \varepsilon_{ij}\, \varepsilon_{kl} > 0, \qquad \sum_{j,l=1}^{3} \epsilon_{jl}\, E_j\, E_l > 0 \tag{2.11}$$

for any non-zero 3×3 real symmetric matrix (ε_{ij}) and for any non-zero real vector $(E_1, E_2, E_3) \in \mathbb{R}^3$.

The equations of mechanical and electric equilibrium with zero body force and zero free charge are given by

$$\sum_{j=1}^{3} \frac{\partial \sigma_{ij}}{\partial x_j} = 0, \qquad i = 1,2,3 \quad \text{and} \quad \sum_{j=1}^{3} \frac{\partial D_j}{\partial x_j} = 0, \tag{2.12}$$

respectively.

To give unified expressions for the constitutive equations (2.9) and for the equilibrium equations (2.12), we introduce the coefficients B_{ijkl} for $i, k = 1, 2, 3, 4$, $j, l = 1, 2, 3$ by

$$
B_{ijkl} = \begin{cases}
C_{ijkl} & i, j, k, l = 1, 2, 3 \\[2mm]
e_{ijl} & i, j, l = 1, 2, 3, \quad k = 4 \\[2mm]
e_{klj} & j, k, l = 1, 2, 3, \quad i = 4 \\[2mm]
-\epsilon_{jl} & j, l = 1, 2, 3, \quad i = k = 4
\end{cases}
\tag{2.13}
$$

and define a four-dimensional vector $(\tilde{u}_k)_{k=1,2,3,4}$ by

$$
\tilde{u}_k = \begin{cases}
u_k & k = 1, 2, 3 \\[2mm]
\phi & k = 4
\end{cases}
\tag{2.14}
$$

and a 4×3 tensor $(\tilde{\sigma}_{ij})_{i=1,2,3,4,\, j=1,2,3}$ by

$$
\tilde{\sigma}_{ij} = \begin{cases}
\sigma_{ij} & i, j = 1, 2, 3 \\[2mm]
D_j & i = 4.
\end{cases}
\tag{2.15}
$$

Then (2.9) is written as

$$
\tilde{\sigma}_{ij} = \sum_{k=1}^{4} \sum_{l=1}^{3} B_{ijkl} \frac{\partial \tilde{u}_k}{\partial x_l}, \qquad i = 1, 2, 3, 4, \qquad j = 1, 2, 3
\tag{2.16}
$$

and (2.12) is written as

$$
\sum_{j=1}^{3} \frac{\partial \tilde{\sigma}_{ij}}{\partial x_j} = 0, \qquad i = 1, 2, 3, 4.
$$

In this section we consider homogeneous piezoelectric materials. Then $\mathbf{C}$, e, and ϵ are independent of $\mathbf{x} = (x_1, x_2, x_3)$, and the equilibrium equations become

$$
\sum_{k=1}^{4} \sum_{j,l=1}^{3} B_{ijkl} \frac{\partial^2 \tilde{u}_k}{\partial x_j \partial x_l} = 0, \qquad i = 1, 2, 3, 4.
\tag{2.17}
$$

Equations (2.16) and (2.17) are parallel to (1.5) and (1.17), respectively, except that the indices i and k run from 1 to 4.

The purpose of this section is to extend the Stroh formalism in Chapter 1 to piezoelectricity.

2.2.2 Extension of the Stroh Formalism

Let $\mathbf{m} = (m_1, m_2, m_3), \mathbf{n} = (n_1, n_2, n_3)$ be orthogonal unit vectors in $\mathbb{R}^3$. Using B_{ijkl} in (2.13), we define 4×4 matrices $\mathbf{Q}, \mathbf{R}$ and $\mathbf{T}$ by

$$\mathbf{Q} = \left(\sum_{j,l=1}^{3} B_{ijkl} \, m_j m_l \right)_{i \downarrow k \to 1,2,3,4} , \quad \mathbf{R} = \left(\sum_{j,l=1}^{3} B_{ijkl} \, m_j n_l \right)_{i \downarrow k \to 1,2,3,4} ,$$

$$\mathbf{T} = \left(\sum_{j,l=1}^{3} B_{ijkl} \, n_j n_l \right)_{i \downarrow k \to 1,2,3,4} . \tag{2.18}$$

Corresponding to Lemma 1.1, we have

Lemma 2.5

(1) *The matrices $\mathbf{Q}$ and $\mathbf{T}$ are symmetric and invertible.*
(2) *The characteristic roots p_α $(1 \le \alpha \le 8)$, i.e., the solutions to the eighth-order equation*

$$\det \left[\mathbf{Q} + p \left(\mathbf{R} + \mathbf{R}^T \right) + p^2 \mathbf{T} \right] = 0 \tag{2.19}$$

are not real and they occur in complex conjugate pairs.

Proof By (2.13), $\left(B_{ijkl} \right)_{i,k=1,2,3,4, \, j,l=1,2,3}$ do not have the minor symmetries any more, but they still enjoy the major symmetry

$$B_{ijkl} = B_{klij}, \qquad i, k = 1, 2, 3, 4, \qquad j, l = 1, 2, 3.$$

Hence the matrices $\mathbf{Q}$ and $\mathbf{T}$ are symmetric. Moreover, we see from (2.13) that the 4×4 matrix $\mathbf{Q}$ can be written blockwise as

$$\mathbf{Q} = \begin{bmatrix} \mathbf{Q}_C & \mathbf{q}_e \\ \mathbf{q}_e^T & -Q_\epsilon \end{bmatrix},$$

where the 3×3 matrix $\mathbf{Q}_C$, the three-dimensional column vector $\mathbf{q}_e$, and the scalar Q_ϵ are given by

$$\mathbf{Q}_C = \left(\sum_{j,l=1}^{3} C_{ijkl} \, m_j m_l \right)_{i \downarrow k \to 1,2,3} , \quad \mathbf{q}_e = \left(\sum_{j,l=1}^{3} e_{ijl} \, m_j m_l \right)_{i \downarrow 1,2,3} ,$$

$$Q_\epsilon = \sum_{j,l=1}^{3} \epsilon_{jl} \, m_j m_l .$$

🍃 Springer

Then it follows that

$$Q = \begin{bmatrix} \mathbf{I} & \mathbf{0} \\ \mathbf{0}^T & -1 \end{bmatrix} \begin{bmatrix} \mathbf{Q}_C & \mathbf{q}_e \\ -\mathbf{q}_e^T & Q_\epsilon \end{bmatrix},$$

where $\mathbf{I}$ is the 3×3 identity matrix and $\mathbf{0}$ is the three-dimensional column zero vector. Since it follows from (2.11) that $\mathbf{Q}_C$ is positive definite (Lemma 1.1) and that Q_ϵ is a positive scalar, we see that the 4×4 matrix

$$\begin{bmatrix} \mathbf{Q}_C & \mathbf{q}_e \\ -\mathbf{q}_e^T & Q_\epsilon \end{bmatrix}$$

is positive definite, and hence the matrix $\mathbf{Q}$ is invertible. In the same way, it is proved that $\mathbf{T}$ is invertible.

The proof of (2) is exactly the same as that of (2) of Lemma 1.1 except that the indices i and k run from 1 to 4 with C_{ijkl} replaced by B_{ijkl} and that the matrix $\left(\sum_{j,l=1}^{3} B_{ijkl} w_j w_l \right)_{i\downarrow k \to 1,2,3,4}$ is invertible. □

Therefore, the arguments in Chapter 1 can be applied to piezoelectricity. Here, the 3×3 real matrices $\mathbf{Q}$, $\mathbf{R}$ and $\mathbf{T}$ in (1.20) are replaced by the 4×4 real matrices $\mathbf{Q}$, $\mathbf{R}$ and $\mathbf{T}$ in (2.18), and the 6×6 real matrix $\mathbf{N}$ in (1.30) is replaced by the 8×8 real matrix $\mathbf{N}$ with $\mathbf{Q}$, $\mathbf{R}$ and $\mathbf{T}$ given by (2.18). Then the eigenvalue problem (1.29) becomes an eight-dimensional eigenrelation; here the vector $\mathbf{a}$ becomes a four-dimensional complex vector whose first three components pertain to the mechanical displacement and whose last component to the electric potential (see (2.14)). Also, the vector $\mathbf{l}$ becomes a four-dimensional complex vector whose first three components pertain to the mechanical traction on the surface $\mathbf{n} \cdot \mathbf{x} = 0$ and whose last component to the normal component of the electric displacement there (consider $\sum_{j=1}^{3} \tilde{\sigma}_{ij} n_j, i = 1, 2, 3, 4$, where $\tilde{\sigma}_{ij}$ are given by (2.15)).

When we consider the dependence of eigenvectors and generalized eigenvectors of this extended eight-dimensional eigenvalue problem on the rotations of $\mathbf{m}$ and $\mathbf{n}$, we take the 4×4 real matrices $\mathbf{Q}(\phi), \mathbf{R}(\phi), \mathbf{T}(\phi)$ and the 8×8 real matrix $\mathbf{N}(\phi)$ to be those which are defined by using $\mathbf{Q}$, $\mathbf{R}$ and $\mathbf{T}$ in (2.18) with the rotated $\mathbf{m} = (m_1, m_2, m_3)$, $\mathbf{n} = (n_1, n_2, n_3)$ given by (1.31) (cf. (1.32) and (1.33)). Using these matrices, we define the 8×8 real matrix $\mathbf{S}$ (i.e., the angular average of $\mathbf{N}(\phi)$) and its 4×4 blocks $\mathbf{S}_i$ ($i = 1, 2, 3$) by (1.67) and (1.68), respectively.

When establishing the Stroh formalism in piezoelectricity, we shall need to consider the case where the eight-dimensional eigenvalue problem (1.29) has a Jordan chain of length 4. When not only the piezoelectric effect but also the piezomagnetic and magnetoelectric effects are present, the corresponding matrices $\mathbf{Q}$, $\mathbf{R}$ and $\mathbf{T}$ become 5×5 matrices and the matrices $\mathbf{N}$ and $\mathbf{S}$ become 10×10 matrices. Then the eigenvalue problem (1.29) becomes a ten-dimensional eigenrelation and there is a possibility that a Jordan chain of length 5 appears.

From now on in this section we prove a higher dimensional version of Theorem 1.15. The proof does not depend on the dimension of the eigenvalue problem (1.29) and is valid whatever length its Jordan chain has. This theorem becomes a basis for the integral formalism in a higher dimension.

Theorem 2.6 *Let* $\begin{bmatrix} \mathbf{a}_1 \\ \mathbf{l}_1 \end{bmatrix}$ *be an eigenvector of* $\mathbf{N}(0)$ *corresponding to an eigenvalue* p_1 *with* $\mathrm{Im}\, p_1 > 0$, *and let* $\begin{bmatrix} \mathbf{a}_\alpha \\ \mathbf{l}_\alpha \end{bmatrix}$ $(\alpha = 2, \cdots, n)$ *be generalized eigenvectors which form a Jordan chain of* $\mathbf{N}(0)$ *of length* n *so that*

$$\mathbf{N}(0) \begin{bmatrix} \mathbf{a}_\alpha \\ \mathbf{l}_\alpha \end{bmatrix} - p_1 \begin{bmatrix} \mathbf{a}_\alpha \\ \mathbf{l}_\alpha \end{bmatrix} = \begin{bmatrix} \mathbf{a}_{\alpha-1} \\ \mathbf{l}_{\alpha-1} \end{bmatrix}, \qquad \alpha = 2, \cdots, n. \tag{2.20}$$

Then it holds that

$$\mathbf{S} \begin{bmatrix} \mathbf{a}_\alpha \\ \mathbf{l}_\alpha \end{bmatrix} = \sqrt{-1} \begin{bmatrix} \mathbf{a}_\alpha \\ \mathbf{l}_\alpha \end{bmatrix}, \qquad \alpha = 1, 2, \cdots, n. \tag{2.21}$$

To prove the theorem, we need three lemmas.

Lemma 2.7

$$\mathbf{N}'(\phi) = -\mathbf{I} - \mathbf{N}(\phi)^2, \tag{2.22}$$

where $\mathbf{I}$ *is the identity matrix the dimension of which is equal to that of* $\mathbf{N}(\phi)$ *and* $' = \frac{d}{d\phi}$.

For the proof, see the comments on (1.144).

Lemma 2.8 *Let* $p(\phi)$ *be the solution to the Riccati equation* (1.34) *with* $p(0) = p_1$. *Moreover, let* $a_1(\phi), \ldots, a_{n-1}(\phi)$ *be the solutions to the following initial value problems:*

$$a_1'(\phi) = -2\, p(\phi)\, a_1(\phi), \qquad a_1(0) = 1, \tag{2.23}$$

$$\begin{cases} a_\alpha'(\phi) = -2p(\phi)a_\alpha(\phi) - \sum_{\beta=1}^{\alpha-1} a_{\alpha-\beta}(\phi)a_\beta(\phi), \\ a_\alpha(0) = 0, \qquad \alpha = 2, \ldots, n-1. \end{cases} \tag{2.24}$$

Then for $\alpha = 2, \ldots, n$, *we have*

$$\left(\mathbf{N}(\phi) - p(\phi)\mathbf{I}\right) \begin{bmatrix} \mathbf{a}_\alpha \\ \mathbf{l}_\alpha \end{bmatrix} = \sum_{\beta=1}^{\alpha-1} a_{\alpha-\beta}(\phi) \begin{bmatrix} \mathbf{a}_\beta \\ \mathbf{l}_\beta \end{bmatrix} \tag{2.25}$$

for all ϕ.

Proof We set

$$\mathbf{f}_\alpha(\phi) = \left(\mathbf{N}(\phi) - p(\phi)\mathbf{I}\right) \begin{bmatrix} \mathbf{a}_\alpha \\ \mathbf{l}_\alpha \end{bmatrix} - \sum_{\beta=1}^{\alpha-1} a_{\alpha-\beta}(\phi) \begin{bmatrix} \mathbf{a}_\beta \\ \mathbf{l}_\beta \end{bmatrix}, \qquad \alpha = 2, \ldots, n. \tag{2.26}$$

From (2.20) it follows that

$$\mathbf{f}_2(0) = \left(\mathbf{N}(0) - p(0)\mathbf{I}\right) \begin{bmatrix} \mathbf{a}_2 \\ \mathbf{l}_2 \end{bmatrix} - \begin{bmatrix} \mathbf{a}_1 \\ \mathbf{l}_1 \end{bmatrix} = \mathbf{0}.$$

🍀 Springer

From Lemma 2.7, (1.34) and (2.23), it follows that

$$\mathbf{f}_2'(\phi) = \left[-\mathbf{N}(\phi)^2 + p(\phi)^2\mathbf{I}\right]\begin{bmatrix}\mathbf{a}_2\\ \mathbf{l}_2\end{bmatrix} + 2p(\phi)\,a_1(\phi)\begin{bmatrix}\mathbf{a}_1\\ \mathbf{l}_1\end{bmatrix},$$

and from (2.26) and Theorem 1.5,

$$\mathbf{f}_2'(\phi) = -\Big(\mathbf{N}(\phi) + p(\phi)\mathbf{I}\Big)\mathbf{f}_2(\phi).$$

Therefore, the uniqueness of the solution to the initial value problem implies that

$$\mathbf{f}_2(\phi) = \mathbf{0}$$

for all ϕ. This proves (2.25) for $\alpha = 2$.

Suppose that (2.25) holds for $\alpha = 2, \ldots, m-1$ ($m \geq 3$). Then by (2.20), (2.23) and (2.24),

$$\mathbf{f}_m(0) = \Big(\mathbf{N}(0) - p(0)\mathbf{I}\Big)\begin{bmatrix}\mathbf{a}_m\\ \mathbf{l}_m\end{bmatrix} - \begin{bmatrix}\mathbf{a}_{m-1}\\ \mathbf{l}_{m-1}\end{bmatrix} = \mathbf{0}.$$

Moreover from Lemma 2.7, (1.34), (2.23), (2.24) and (2.26), it follows that

$$\mathbf{f}_m'(\phi) = -\Big(\mathbf{N}(\phi) + p(\phi)\mathbf{I}\Big)\Big(\mathbf{N}(\phi) - p(\phi)\mathbf{I}\Big)\begin{bmatrix}\mathbf{a}_m\\ \mathbf{l}_m\end{bmatrix} - \sum_{\beta=1}^{m-1} a_{m-\beta}'(\phi)\begin{bmatrix}\mathbf{a}_\beta\\ \mathbf{l}_\beta\end{bmatrix}$$

$$= -\Big(\mathbf{N}(\phi) + p(\phi)\mathbf{I}\Big)\left(\mathbf{f}_m(\phi) + \sum_{\beta=1}^{m-1} a_{m-\beta}(\phi)\begin{bmatrix}\mathbf{a}_\beta\\ \mathbf{l}_\beta\end{bmatrix}\right)$$

$$+ \sum_{\beta=1}^{m-2}\left(2p(\phi)a_{m-\beta}(\phi) + \sum_{\gamma=1}^{m-\beta-1} a_{m-\beta-\gamma}(\phi)a_\gamma(\phi)\right)\begin{bmatrix}\mathbf{a}_\beta\\ \mathbf{l}_\beta\end{bmatrix}$$

$$+ 2p(\phi)a_1(\phi)\begin{bmatrix}\mathbf{a}_{m-1}\\ \mathbf{l}_{m-1}\end{bmatrix},$$

and regrouping gives

$$\mathbf{f}_m'(\phi) = -\Big(\mathbf{N}(\phi) + p(\phi)\mathbf{I}\Big)\mathbf{f}_m(\phi) - \sum_{\beta=1}^{m-1} a_{m-\beta}(\phi)\Big(\mathbf{N}(\phi) - p(\phi)\mathbf{I}\Big)\begin{bmatrix}\mathbf{a}_\beta\\ \mathbf{l}_\beta\end{bmatrix}$$

$$+ \sum_{\beta=1}^{m-2}\sum_{\gamma=1}^{m-\beta-1} a_{m-\beta-\gamma}(\phi)a_\gamma(\phi)\begin{bmatrix}\mathbf{a}_\beta\\ \mathbf{l}_\beta\end{bmatrix},$$

and the hypothesis of induction leads us to

$$\mathbf{f}_m'(\phi) = -\Big(\mathbf{N}(\phi) + p(\phi)\mathbf{I}\Big)\mathbf{f}_m(\phi) - \sum_{\beta=2}^{m-1}\sum_{\gamma=1}^{\beta-1} a_{m-\beta}(\phi)a_{\beta-\gamma}(\phi)\begin{bmatrix}\mathbf{a}_\gamma\\ \mathbf{l}_\gamma\end{bmatrix}$$

$$+ \sum_{\beta=1}^{m-2}\sum_{\gamma=1}^{m-\beta-1} a_{m-\beta-\gamma}(\phi)a_\gamma(\phi)\begin{bmatrix}\mathbf{a}_\beta\\ \mathbf{l}_\beta\end{bmatrix},$$

where Theorem 1.5 has been used for $\beta = 1$ in the second term. Since the last two terms cancel each other out by changing the order of summation in the second term

$$\sum_{\beta=2}^{m-1}\sum_{\gamma=1}^{\beta-1} = \sum_{\gamma=1}^{m-2}\sum_{\beta=\gamma+1}^{m-1},$$

we have

$$\mathbf{f}'_m(\phi) = -\Big(\mathbf{N}(\phi) + p(\phi)\mathbf{I}\Big)\mathbf{f}_m(\phi).$$

Thus, from the uniqueness of the solution to the initial value problem we obtain

$$\mathbf{f}_m(\phi) = \mathbf{0}$$

for all ϕ. $\qquad\square$

Now we set

$$q(\phi) = \exp\left(-\int_0^{\phi} 2p(\psi)d\psi\right), \qquad r(\phi) = \int_0^{\phi} q(\psi)d\psi. \tag{2.27}$$

Lemma 2.9 *The solutions* $a_1(\phi), \ldots, a_{n-1}(\phi)$ *to* (2.23) *and* (2.24) *have the following form:*

$$a_\alpha(\phi) = (-1)^{\alpha-1} r(\phi)^{\alpha-1} r'(\phi), \qquad \alpha = 1, \ldots, n-1. \tag{2.28}$$

Proof Integrating (2.23), we get (2.28) for $\alpha = 1$. Suppose that (2.28) holds for $\alpha = 1, \ldots, m-1$. Solving (2.24), we have for $m \geq 2$,

$$a_m(\phi) = -q(\phi)\int_0^{\phi} q(\psi)^{-1}\sum_{\beta=1}^{m-1} a_\beta(\psi)a_{m-\beta}(\psi)d\psi. \tag{2.29}$$

From the hypothesis of induction and $r'(\psi) = q(\psi)$, it follows that

$$q(\psi)^{-1}a_\beta(\psi)a_{m-\beta}(\psi) = (-1)^{m-2}r(\psi)^{m-2}r'(\psi).$$

Thus

$$a_m(\phi) = -q(\phi)\int_0^{\phi} (m-1)(-1)^{m-2}r(\psi)^{m-2}r'(\psi)d\psi = (-1)^{m-1}r(\phi)^{m-1}r'(\phi).$$

$\qquad\square$

Proof of Theorem 2.6 We take the angular average of both sides of (2.25) over $[-\pi, \pi]$. By Lemma 1.12, the angular average of the left hand side becomes

$$\Big(\mathbf{S} - \sqrt{-1}\,\mathbf{I}\Big)\begin{bmatrix}\mathbf{a}_\alpha \\ \mathbf{l}_\alpha\end{bmatrix}.$$

From (1.71) and (1.72), it follows that $r(\pi) = r(-\pi)$. Then Lemma 2.9 implies that

$$\int_{-\pi}^{\pi} a_{\alpha-\beta}(\phi)d\phi = (-1)^{\alpha-\beta-1}\left[\frac{r(\phi)^{\alpha-\beta}}{\alpha-\beta}\right]_{\phi=-\pi}^{\phi=\pi} = 0, \qquad \beta = 1, 2, \ldots, \alpha-1.$$

This proves the theorem. $\qquad\square$

2.2.3 Surface Impedance Tensor of Piezoelectricity

Let $\mathbf{m}$ and $\mathbf{n}$ be two orthogonal unit vectors in $\mathbb{R}^3$, and let $\begin{bmatrix} \mathbf{a}_\alpha \\ \mathbf{l}_\alpha \end{bmatrix}$ $(\alpha = 1, 2, 3, 4)$ be linearly independent eigenvector(s) or generalized eigenvector(s) of the eight-dimensional eigenvalue problem (1.29) associated with the eigenvalues p_α $(\alpha = 1, 2, 3, 4,\ \mathrm{Im}\, p_\alpha > 0)$. Here $\mathbf{Q}, \mathbf{R}$ and $\mathbf{T}$ are taken to be the 4×4 matrices in (2.18). Corresponding to Theorem 1.16, we obtain

Theorem 2.10 *The vectors* $\mathbf{a}_\alpha$ $(\alpha = 1, 2, 3, 4)$ *are linearly independent.*

Proof On the basis of Theorem 2.6, we can use the proof of Theorem 1.16. We only have to change the range where the index α runs from $\{1, 2, 3\}$ to $\{1, 2, 3, 4\}$. To use the proof of Theorem 1.16 with this new index, it is sufficient to prove the invertibility of the 4×4 matrix $\mathbf{S}_2$ defined by (1.68) with the 4×4 matrix $\mathbf{T}(\phi)$ given by (2.18) and (1.31). As in the proof of (1) of Lemma 2.5, we can write

$$\mathbf{T}(\phi) = \begin{bmatrix} \mathbf{I} & \mathbf{0} \\ \mathbf{0}^T & -1 \end{bmatrix} \begin{bmatrix} \mathbf{T}_C(\phi) & \mathbf{t}_e(\phi) \\ -\mathbf{t}_e^T(\phi) & \mathbf{T}_\epsilon(\phi) \end{bmatrix},$$

where

$$\mathbf{T}_C(\phi) = \left(\sum_{j,l=1}^{3} C_{ijkl}\, n_j n_l \right)_{i\downarrow k\to 1,2,3} \quad , \quad \mathbf{t}_e(\phi) = \left(\sum_{j,l=1}^{3} e_{ijl}\, n_j n_l \right)_{i\downarrow 1,2,3} ,$$

$$\mathbf{T}_\epsilon(\phi) = \sum_{j,l=1}^{3} \epsilon_{jl}\, n_j n_l$$

and $\mathbf{n} = (n_1, n_2, n_3)$ is defined in (1.31). Then it follows that

$$\mathbf{S}_2 = \frac{1}{2\pi} \int_{-\pi}^{\pi} \begin{bmatrix} \mathbf{T}_C(\phi) & \mathbf{t}_e(\phi) \\ -\mathbf{t}_e^T(\phi) & \mathbf{T}_\epsilon(\phi) \end{bmatrix}^{-1} d\phi \begin{bmatrix} \mathbf{I} & \mathbf{0} \\ \mathbf{0}^T & -1 \end{bmatrix}.$$

By (2.11), the matrix $\begin{bmatrix} \mathbf{T}_C(\phi) & \mathbf{t}_e(\phi) \\ -\mathbf{t}_e^T(\phi) & \mathbf{T}_\epsilon(\phi) \end{bmatrix}$ and hence its inverse are positive definite for all ϕ. Therefore, the angular average on the right hand side is positive definite. Since $\begin{bmatrix} \mathbf{I} & \mathbf{0} \\ \mathbf{0}^T & -1 \end{bmatrix}$ is invertible, $\mathbf{S}_2$ is also invertible. □

Thus, we can define the surface impedance tensor of piezoelectricity.

Definition 2.11 The surface impedance tensor $\mathbf{Z}$ of piezoelectricity is the 4×4 matrix given by

$$\mathbf{Z} = -\sqrt{-1}\, [\mathbf{l}_1, \mathbf{l}_2, \mathbf{l}_3, \mathbf{l}_4][\mathbf{a}_1, \mathbf{a}_2, \mathbf{a}_3, \mathbf{a}_4]^{-1}, \tag{2.30}$$

where $[\mathbf{l}_1, \mathbf{l}_2, \mathbf{l}_3, \mathbf{l}_4]$ and $[\mathbf{a}_1, \mathbf{a}_2, \mathbf{a}_3, \mathbf{a}_4]$ denote 4×4 matrices which consist of the column vectors $\mathbf{l}_\alpha$ and $\mathbf{a}_\alpha$, respectively.

We also have the integral representation of $\mathbf{Z}$.

Theorem 2.12

$$\mathbf{Z} = \mathbf{S}_2^{-1} + \sqrt{-1}\,\mathbf{S}_2^{-1}\mathbf{S}_1, \tag{2.31}$$

where the 4×4 matrices $\mathbf{S}_1$ and $\mathbf{S}_2$ are defined by (1.68) with the 4×4 matrices $\mathbf{Q}(\phi)$, $\mathbf{R}(\phi)$ and $\mathbf{T}(\phi)$ given by (2.18) and (1.31).

Then the arguments in Subsection 2.1.1 can be applied to obtain

Theorem 2.13

$$\mathbf{G}(\mathbf{x}) = \frac{1}{4\pi\,|\mathbf{x}|}\,(\mathrm{Re}\,\mathbf{Z})^{-1} \tag{2.32}$$

is a fundamental solution for piezoelectricity, where $\mathbf{Z}$ is the surface impedance tensor defined by (2.30) with $\mathbf{m}$ and $\mathbf{n}$ satisfying

$$\mathbf{m} \times \mathbf{n} = \frac{\mathbf{x}}{|\mathbf{x}|}.$$

2.2.4 Formula for Surface Impedance Tensor of Piezoelectricity: Example

Using the algebraic formula (2.30) for the surface impedance tensor, we give the explicit form of $\mathbf{Z}$ for piezoelectric materials which have hexagonal symmetry 622. For such materials a Cartesian coordinate system can be chosen so that one coordinate axis is an axis of 6-fold rotational symmetry and the other two axes are 2-fold. Henceforth, we take the 3-axis to be the 6-fold axis and the 1- and 2- axes to be the 2-fold axes. Then the orthogonal tensors $\mathbf{R}_{x_3}^{\frac{\pi}{3}}$ and $\mathbf{R}_{x_2}^{\pi}$ are the generators of the symmetry group $\mathcal{G}$ for this piezoelectric material. It follows that for any orthogonal tensor $\mathbf{Q} = (Q_{ij})_{i,j=1,2,3}$ in $\mathcal{G}$, (1.10) holds for the elasticity tensor $\mathbf{C} = (C_{ijkl})_{i,j,k,l=1,2,3}$,

$$e_{ijl} = \sum_{p,q,r=1}^{3} Q_{ip}\,Q_{jq}\,Q_{lr}\,e_{pqr}, \qquad i,j,l = 1,2,3$$

holds for the piezoelectric tensor $e = (e_{ijl})_{i,j,l=1,2,3}$, and

$$\epsilon_{jl} = \sum_{p,q=1}^{3} Q_{jp}\,Q_{lq}\,\epsilon_{pq}, \qquad j,l = 1,2,3$$

holds for the dielectric tensor $\epsilon = (\epsilon_{jl})_{i,l=1,2,3}$. Then the elasticity tensor $\mathbf{C}$ is given by (1.13), while the non-zero components of the piezoelectric tensor e and the dielectric tensor ϵ are proved to be

$$e_{231},\ e_{132},\ \epsilon_{11},\ \epsilon_{22},\ \epsilon_{33},$$

and they satisfy

$$e_{231} = -e_{132}, \qquad \epsilon_{11} = \epsilon_{22}.$$

🍃 Springer

Hereafter, for simplicity, we use the notations (1.100) and put

$$e_{231} = S, \qquad \epsilon_{11} = V, \qquad \epsilon_{33} = W.$$

For the orthogonal unit vectors $\mathbf{m}$ and $\mathbf{n}$ in $\mathbb{R}^3$, we write $\mathbf{m} \times \mathbf{n} = \boldsymbol{\ell} = (\ell_1, \ell_2, \ell_3)$.

Theorem 2.14 *The surface impedance tensor for piezoelectric materials of hexagonal symmetry 622 is given by*

$$\mathbf{Z} = (Z_{ij})_{i\downarrow j \to 1,2,3,4} = \overline{\mathbf{Z}}^T,$$

$$Z_{11} = \frac{L}{D}\left\{(A-N)M\Theta\ell_1^2 + AG\Phi\ell_2^2\right\},$$

$$Z_{22} = \frac{L}{D}\left\{(A-N)M\Theta\ell_2^2 + AG\Phi\ell_1^2\right\},$$

$$Z_{33} = \frac{1}{D}\left[ALGH\Phi - (A-N)M\left\{AL\left(\ell_3^2 + H\right)\right.\right.$$
$$\left.\left. + \left(AC - (F+L)^2\right)\left(\ell_1^2 + \ell_2^2\right)\right\}\ell_3^2\right],$$

$$Z_{44} = \frac{1}{D}\left[-(A-N)VJM\Theta + AG\left\{(A-N)V\left(\ell_3^2 + J\right)\right.\right.$$
$$\left.\left. + \left((A-N)W + 2S^2\right)\left(\ell_1^2 + \ell_2^2\right)\right\}\ell_3^2\right],$$

$$Z_{12} = \frac{L}{D}\left\{(A-N)M\Theta - AG\Phi\right\}\ell_1\ell_2$$
$$+ \sqrt{-1}(A-N)\left\{-1 + \frac{ALGM}{D}\left(\ell_1^2 + \ell_2^2\right)\right\}\ell_3,$$

$$Z_{13} = \frac{-L(F+L)(A-N)M}{D}\left(\ell_1^2 + \ell_2^2\right)\ell_1\ell_3$$
$$- \sqrt{-1}L\left\{-1 + \frac{(F+L)}{D}\Phi\left(\ell_1^2 + \ell_2^2\right)\right\}\ell_2,$$

$$Z_{23} = \frac{-L(F+L)(A-N)M}{D}\left(\ell_1^2 + \ell_2^2\right)\ell_2\ell_3$$
$$+ \sqrt{-1}L\left\{-1 + \frac{(F+L)}{D}\Phi\left(\ell_1^2 + \ell_2^2\right)\right\}\ell_1,$$

$$Z_{14} = \frac{2ALSG}{D}\left(\ell_1^2 + \ell_2^2\right)\ell_2\ell_3 + \sqrt{-1}S\left\{1 - \frac{2L}{D}\Theta\left(\ell_1^2 + \ell_2^2\right)\right\}\ell_1,$$

$$Z_{24} = -\frac{2ALSG}{D}\left(\ell_1^2 + \ell_2^2\right)\ell_1\ell_3 + \sqrt{-1}S\left\{1 - \frac{2L}{D}\Theta\left(\ell_1^2 + \ell_2^2\right)\right\}\ell_2,$$

$$Z_{34} = \sqrt{-1}S\left\{-1 + \frac{2L(F+L)}{D}\left(\ell_1^2 + \ell_2^2\right)^2\right\}\ell_3, \qquad (2.33)$$

where G and H are the same as those in (1.104) *and*

$$\Gamma = (A - N)V\ell_3{}^4 + ((A - N)W + 2S^2 + 2LV)\left(\ell_1{}^2 + \ell_2{}^2\right)\ell_3{}^2$$
$$+ 2LW\left(\ell_1{}^2 + \ell_2{}^2\right)^2,$$

$$J = \sqrt{\frac{\Gamma}{(A - N)V}},$$

$$M = \sqrt{\frac{2(A - N)V\ell_3{}^4 + ((A - N)W + 2S^2 + 2LV)\left(\ell_1{}^2 + \ell_2{}^2\right)}{(A - N)V}}$$
$$\overline{+2\sqrt{(A - N)V\Gamma}},$$

$$\Theta = A\left(\ell_3{}^2 + H\right) + L\left(\ell_1{}^2 + \ell_2{}^2\right),$$
$$\Phi = (A - N)\left(\ell_3{}^2 + J\right) + 2L\left(\ell_1{}^2 + \ell_2{}^2\right),$$
$$D = \Theta\,\Phi - A(A - N)GM\ell_3{}^2.$$

To obtain the formula above, it is important to see that the characteristic equation (2.19) can be factorized into

$$\frac{1}{2}\left(AL\left(y_1^2 + y_2^2\right)^2 + (AC - F^2 - 2FL)\left(y_1^2 + y_2^2\right)y_3^2 + CLy_3^4\right)$$
$$\times \left((A-N)V\left(y_1^2 + y_2^2\right)^2 + ((A - N)W + 2S^2 + 2LV)\left(y_1^2 + y_2^2\right)y_3^2 + 2LWy_3^4\right),$$

$$(2.34)$$

where $y_i = m_i + p\,n_i$ $(i = 1, 2, 3)$. We note that the first factor of (2.34) is equal to the quartic factor (1.112) of the characteristic equation for transversely isotropic elastic materials, while the second factor of (2.34) depends not only on the elasticity tensor $\mathbf{C}$ but also on the piezoelectric tensor e and the dielectric tensor ϵ. However, for piezoelectric materials which has hexagonal symmetry 622, it is proved that the conditions for the eight-dimensional eigenvalue problem (1.29) to be simple, semisimple and degenerate are precisely the same as those in Lemma 1.25. Hence the arguments in Subsection 1.8.2 applies in an almost parallel way. For the details of computations, we refer to [2].

When

$$e_{231} = S = 0,$$

equations (2.17) can be decomposed into the equilibrium equations for transversely isotropic elasticity and the scalar equation of dielectricity. In this case it can be checked by simple computations that the submatrix $\left(Z_{ij}\right)_{i\downarrow j \rightarrow 1,2,3}$ of the matrix (2.33) is reduced to the matrix (1.103) in Theorem 1.26.

The formula for the fundamental solution can be obtained from Theorem 2.13.

Springer

2.3 Inverse Boundary Value Problem

2.3.1 Dirichlet to Neumann Map

Let Ω be a bounded domain in $\mathbb{R}^3$ occupied by an inhomogeneous, linearly elastic medium. When the displacement $\boldsymbol{f} = (f_1, f_2, f_3)$ is given on the boundary $\partial\Omega$, the deformation of the elastic body due to $\boldsymbol{f}$ can be expressed in terms of the displacement $\boldsymbol{u} = \boldsymbol{u}(\mathbf{x}) = (u_1, u_2, u_3)$ as

$$\sum_{j,k,l=1}^{3} \frac{\partial}{\partial x_j} \left(C_{ijkl} \frac{\partial u_k}{\partial x_l} \right) = 0, \qquad i = 1, 2, 3 \quad \text{in } \Omega, \tag{2.35}$$

$$\boldsymbol{u}\Big|_{\partial\Omega} = \boldsymbol{f}. \tag{2.36}$$

Here $\mathbf{C}(\mathbf{x}) = \left(C_{ijkl} \right)_{i,j,k,l=1,2,3}$ is the elasticity tensor[21] and we assume that this tensor now depends on $\mathbf{x} \in \overline{\Omega}$.

Using the solution $\boldsymbol{u}$ to (2.35) and (2.36), we define the traction $\boldsymbol{t}$ on the boundary $\partial\Omega$ by

$$\boldsymbol{t} = \left(\sum_{j,k,l=1}^{3} C_{ijkl} \frac{\partial u_k}{\partial x_l} n_j \right)_{i\downarrow 1,2,3} \Bigg|_{\partial\Omega}, \tag{2.37}$$

where $\mathbf{n} = \mathbf{n}(\mathbf{x}) = (n_1, n_2, n_3)$ is the unit outward normal to $\partial\Omega$ at $\mathbf{x} \in \partial\Omega$.

The boundary displacement $\boldsymbol{f}$ in (2.36) is the Dirichlet data, while the traction $\boldsymbol{t}$ in (2.37) is the Neumann data. We call the map

$$\Lambda : \boldsymbol{f} \longmapsto \boldsymbol{t} \tag{2.38}$$

the Dirichlet to Neumann map. Physically, $\Lambda(\boldsymbol{f})$ gives the traction on $\partial\Omega$ produced by the displacement $\boldsymbol{f}$ on $\partial\Omega$.

When $\mathbf{C}(\mathbf{x})$ is given in $\overline{\Omega}$ and $\boldsymbol{f}$ is prescribed on $\partial\Omega$, the problem of finding the solution $\boldsymbol{u}$ to (2.35) and (2.36) is the forward problem.

On the other hand, suppose that the elasticity tensor $\mathbf{C}(\mathbf{x})$ is unknown. In this section we are concerned with the following inverse problem, which has received a great deal of attention:

Determine the elasticity tensor $\mathbf{C}(\mathbf{x})$ from the Dirichlet to Neumann map Λ.

This is the problem of determining the elasticity tensor from observations of the displacements on the boundary of the medium and the tractions needed to sustain them on that boundary. Inverse problems where observations are made on the boundary like this are called inverse boundary value problems.

---

[21] Hence $\mathbf{C}(\mathbf{x}) = \left(C_{ijkl} \right)_{i,j,k,l=1,2,3}$ satisfies the symmetry conditions (1.4), (1.6) and the strong convexity condition (1.7) at each $\mathbf{x} \in \overline{\Omega}$.

We formulate the Dirichlet to Neumann map more precisely using the function spaces. Let $\Omega \in \mathbb{R}^3$ be a bounded domain with a Lipschitz boundary $\partial\Omega$ [22] and let $\mathbf{C}(\mathbf{x}) = \left(C_{ijkl}\right)_{i,j,k,l=1,2,3} \in L^\infty(\Omega)$, $\boldsymbol{f} \in H^{\frac{1}{2}}(\partial\Omega)$. Then the Dirichlet problem (2.35), (2.36) admits a unique weak solution $\boldsymbol{u}$ in $H^1(\Omega)$.[23]

Definition 2.15 The Dirichlet to Neumann map Λ

$$H^{\frac{1}{2}}(\partial\Omega) \ni \boldsymbol{f} \longmapsto \Lambda(\boldsymbol{f}) \in H^{\frac{-1}{2}}(\partial\Omega)$$

is defined by

$$< \Lambda(\boldsymbol{f}), \boldsymbol{g} > = \int_\Omega \sum_{i,j,k,l=1}^3 C_{ijkl} \frac{\partial u_k}{\partial x_l} \frac{\partial v_i}{\partial x_j} \, d\mathbf{x} \tag{2.39}$$

for $\boldsymbol{g} = (g_1, g_2, g_3) \in H^{1/2}(\partial\Omega)$. Here $\boldsymbol{u} = (u_1, u_2, u_3)$ is the solution to (2.35) and (2.36), $\boldsymbol{v} = (v_1, v_2, v_3)$ is any function in $H^1(\Omega)$ satisfying $\boldsymbol{v}|_{\partial\Omega} = \boldsymbol{g}$ and $< , >$ is the bilinear pairing between $H^{1/2}(\partial\Omega)$ and $H^{-1/2}(\partial\Omega)$.

We note that the right hand side of (2.39) is independent of the choice of the extension $\boldsymbol{v} \in H^1(\Omega)$ of $\boldsymbol{g}$. In fact, for another extension $\tilde{\boldsymbol{v}} = (\tilde{v}_1, \tilde{v}_2, \tilde{v}_3)$ of $\boldsymbol{g}$ we have $\boldsymbol{v} - \tilde{\boldsymbol{v}} \in H_0^1(\Omega)$. Then, by the divergence theorem, the difference between the corresponding right hand sides becomes

$$\int_\Omega \sum_{i,j,k,l=1}^3 C_{ijkl} \frac{\partial u_k}{\partial x_l} \frac{\partial(v_i - \tilde{v}_i)}{\partial x_j} \, d\mathbf{x}$$

$$= -\int_\Omega \sum_{i,j,k,l=1}^3 \frac{\partial}{\partial x_j} \left(C_{ijkl} \frac{\partial u_k}{\partial x_l} \right) (v_i - \tilde{v}_i) \, d\mathbf{x}, \text{[24]}$$

which, by (2.35), is equal to zero. Hence (2.39) is well-defined.

---

[22] Ω is said to have a Lipschitz boundary $\partial\Omega$ if $\partial\Omega$ can be viewed locally as the graph of a Lipschitz function after an appropriate rotation of the coordinate axes. In the following we will give the function spaces which will be needed in this section.

$L^\infty(\Omega)$: space of measurable functions in Ω which are essentially bounded.

$H^1(\Omega)$: space of functions $\boldsymbol{u}$ in Ω such that $\frac{\partial^\alpha \boldsymbol{u}}{\partial \mathbf{x}^\alpha} \in L^2(\Omega)$ for all multi-index α with $|\alpha| \le 1$.

$H_0^1(\Omega)$: space of functions $\boldsymbol{u} \in H^1(\Omega)$ such that $\boldsymbol{u}|_{\partial\Omega} = \boldsymbol{0}$. This space is equal to the closure in $H^1(\Omega)$ of $C_c^\infty(\Omega)$ with the norm $\|\boldsymbol{u}\|_{H^1(\Omega)} = \sum_{|\alpha| \le 1} \|\frac{\partial^\alpha \boldsymbol{u}}{\partial \mathbf{x}^\alpha}\|_{L^2(\Omega)}$, where $C_c^\infty(\Omega)$ is the space of C^∞ functions with compact support in Ω.

$H^{\frac{1}{2}}(\partial\Omega)$: the trace of $\boldsymbol{u} \in H^1(\Omega)$ on $\partial\Omega$.

$H^{-1}(\Omega)$: the dual of $H_0^1(\Omega)$.

$H^{-\frac{1}{2}}(\partial\Omega)$: the dual of $H^{\frac{1}{2}}(\partial\Omega)$.

[23] This is proved by Korn's inequality which is derived from the convexity condition (1.7) and by the Lax-Milgram theorem. For the details we refer to [26, 28, 30].

[24] Precisely, since $\boldsymbol{v} - \tilde{\boldsymbol{v}} \in H_0^1(\Omega)$, there exists a sequence $\{\boldsymbol{\phi}_n\}_{n=1}^\infty \subset C_c^\infty(\Omega)$ such that $\boldsymbol{\phi}_n \longrightarrow \boldsymbol{v} - \tilde{\boldsymbol{v}}$ in $H^1(\Omega)$ as $n \longrightarrow \infty$. Hence this right hand side means

$$-\lim_{n\to\infty} \int_\Omega \sum_{i,j,k,l=1}^3 \frac{\partial}{\partial x_j} \left(C_{ijkl} \frac{\partial u_k}{\partial x_l} \right) (\boldsymbol{\phi}_n)_i \, d\mathbf{x},$$

Springer

When f, g, $\mathbf{C}(\mathbf{x})$ and the boundary $\partial\Omega$ have enough regularity, from the divergence theorem and (2.35) again, it follows that

$$< \Lambda(f),\ g > = \int_{\partial\Omega} \sum_{i,j,k,l=1}^{3} C_{ijkl}(\mathbf{x}) \frac{\partial u_k}{\partial x_l}\, n_j v_i\, d\sigma$$

$$= \int_{\partial\Omega} t(\mathbf{x}) \cdot g\, d\sigma,$$

where $d\sigma$ denotes the surface element on $\partial\Omega$. This shows that (2.39) provides a weak formulation of (2.38).

In this section we consider the problem of reconstructing elasticity tensors $\mathbf{C}(\mathbf{x})$ at the boundary $\partial\Omega$ from the localized Dirichlet to Neumann map for isotropic and for transversely isotropic materials.

2.3.2 Reconstruction of Elasticity Tensor

Suppose that the elasticity tensor $\mathbf{C}(\mathbf{x}) = \left(C_{ijkl}\right)_{i,j,k,l=1,2,3} \in L^\infty(\Omega)$ is unknown. We consider the problem of reconstructing $\mathbf{C}(\mathbf{x})$ on the boundary $\partial\Omega$ from the localized Dirichlet to Neumann map. More precisely, for $\mathbf{x}_0 \in \partial\Omega$, assuming that $\partial\Omega$ is locally C^1 near $\mathbf{x}_0$ and that $\mathbf{C}(\mathbf{x})$ is continuous around $\mathbf{x}_0$, we input the Dirichlet data f compactly supported in a small neighborhood of $\mathbf{x}_0$ on $\partial\Omega$, measure the corresponding Neumann data $\Lambda(f)$ in that neighborhood, and from these observations we reconstruct $\mathbf{C}(\mathbf{x}_0)$. This reconstruction problem can be paraphrased as "local reconstruction from local observations".

This reconstruction can be achieved through the following two steps. First, from a countable pairing of f and $\Lambda(f)$ we reconstruct the surface impedance tensor $\mathbf{Z}$ in Section 1.7. Secondly, we reconstruct the components of $\mathbf{C}(\mathbf{x}_0)$ from $\mathbf{Z}$. As $\mathbf{Z}$ expresses a linear relationship between a certain class of displacements given at the surface (Dirichlet data) and the tractions needed to sustain them at that surface (Neumann data) (see the paragraph after Definition 1.17), it is reasonable that $\mathbf{Z}$ can be reconstructed from f and $\Lambda(f)$. The explicit forms of $\mathbf{Z}$ have been written in terms of the elasticity tensors for isotropic and for transversely isotropic materials in Section 1.8. Hence in this subsection we consider the problem of reconstructing $\mathbf{C}(\mathbf{x}_0)$ of isotropic and of transversely isotropic materials.

2.3.2.1 Reconstruction of Surface Impedance Tensor from Localized Dirichlet to Neumann Map

For simplicity we set $\mathbf{x}_0 = \mathbf{0}$. Let $\mathbf{n} = (n_1, n_2, n_3)$ be the unit outward normal to $\partial\Omega$ at $\mathbf{x} = \mathbf{0} \in \partial\Omega$ and we choose an arbitrary unit tangent $\mathbf{m} = (m_1, m_2, m_3)$ to $\partial\Omega$ at $\mathbf{x} = \mathbf{0}$ so that

$$\mathbf{m} \cdot \mathbf{n} = m_1 n_1 + m_2 n_2 + m_3 n_3 = 0.$$

We write

$$\mathbf{m} \times \mathbf{n} = \boldsymbol{\ell} = (\ell_1, \ell_2, \ell_3),$$

---

where $\frac{\partial}{\partial x_j}$ denotes differentiation in the sense of distributions.

where $\mathbf{m} \times \mathbf{n}$ denotes the vector product. For $\mathbf{x} = (x_1, x_2, x_3) \in \overline{\Omega}$ we put

$$y_1 = \mathbf{m} \cdot \mathbf{x}, \qquad y_2 = \boldsymbol{\ell} \cdot \mathbf{x}. \tag{2.40}$$

Since we have assumed that $\partial\Omega$ is locally C^1 near $\mathbf{x}_0$, there exists a C^1 function $\phi(y_1, y_2)$ near $\mathbf{x} = \mathbf{0}$ such that Ω, $\partial\Omega$ are given by

$$\partial\Omega = \{\mathbf{n} \cdot \mathbf{x} = \phi(y_1, y_2)\}, \qquad \Omega = \{\mathbf{n} \cdot \mathbf{x} < \phi(y_1, y_2)\}$$

locally around $\mathbf{x} = \mathbf{0}$ and

$$\frac{\partial\phi}{\partial y_1} = \phi_{y_1} = 0, \qquad \frac{\partial\phi}{\partial y_2} = \phi_{y_2} = 0 \qquad \text{at } \mathbf{x} = \mathbf{0}. \tag{2.41}$$

We set

$$y_3 = \mathbf{n} \cdot \mathbf{x} - \phi(y_1, y_2) \tag{2.42}$$

and introduce the new coordinates

$$\mathbf{y} = (y_1, y_2, y_3) = (y', y_3).$$

Then Ω, $\partial\Omega$ are given by

$$\partial\Omega = \{y_3 = 0\}, \qquad \Omega = \{y_3 < 0\} \tag{2.43}$$

locally around $\mathbf{x} = \mathbf{0}$.

Now we define the Dirichlet data compactly supported in a small neighborhood of $\mathbf{x} = \mathbf{0}$ on $\partial\Omega$. Let $\eta(y')$ be a C^1 function with compact support in $\mathbb{R}^2$ which satisfies

$$0 \le \eta \le 1, \qquad \int_{\mathbb{R}^2} \eta(y')^2 \, dy' = 1, \qquad \operatorname{supp}\eta \subset \{|y'| < 1\}. \tag{2.44}$$

For any vector $\mathbf{v}$ in $\mathbb{C}^3$, define the function $\boldsymbol{f}_N$ on $\partial\Omega$ by

$$\boldsymbol{f}_N = \eta(\sqrt{N}y') \, \mathbf{v} \, e^{-\sqrt{-1}\,y_1 N}\big|_{\partial\Omega}, \tag{2.45}$$

where N is a natural number which is large enough.[25]

The function $\boldsymbol{f}_N$ is localized around $\mathbf{x} = \mathbf{0}$; its support becomes smaller and it oscillates more intensely as N becomes larger. We take $\boldsymbol{f}_N$ as the Dirichlet data, which we use to reconstruct the surface impedance tensor $\mathbf{Z}$ at $\mathbf{x} = \mathbf{0}$.

Theorem 2.16 *Assume that $\partial\Omega$ is locally C^1 at $\mathbf{x} = \mathbf{0} \in \partial\Omega$ and that $\mathbf{C}(\mathbf{x})$ is continuous around $\mathbf{x} = \mathbf{0}$. Letting $\mathbf{v}$ be an arbitrary vector in $\mathbb{C}^3$, we define $\boldsymbol{f}_N$ by (2.45). Then,*

$$\lim_{N \to \infty} < \Lambda(\boldsymbol{f}_N), \overline{\boldsymbol{f}_N} > = (\mathbf{Z}_0 \, \mathbf{v}, \mathbf{v})_{\mathbb{C}^3}, \tag{2.46}$$

where $(\,,\,)_{\mathbb{C}^3}$ denotes the inner product of two vectors in $\mathbb{C}^3$

$$(\mathbf{Z}_0 \, \mathbf{v}, \mathbf{v})_{\mathbb{C}^3} = \sum_{i,j=1}^{3} Z_{ij} v_j \overline{v_i}$$

---

[25] In the exponent, a parameter with the dimension of the reciprocal of length has been set equal to 1 (see (1.18) and its footnote). Hence N can be interpreted as the number of units with the dimension of the reciprocal of length.

and $\mathbf{Z}_0 = (Z_{ij})_{i,j=1,2,3}$ is the surface impedance tensor defined by (1.74) with $\mathbf{m}$ and $\mathbf{n}$ given above and with $\mathbf{C} = (C_{ijkl})_{i,j,k,l=1,2,3} = \mathbf{C}(0)$.

By this theorem, all the components of $\mathbf{Z}_0 = (Z_{ij})_{i,j=1,2,3}$ can be reconstructed from Λ and a suitable choice of $\mathbf{v}$'s in (2.45).

Proof Depending on $\mathbf{C}(0)$ and vectors $\mathbf{m}$ and $\mathbf{n}$, the eigenvalue problem (1.29) becomes simple, semisimple, or degenerate, and when it is degenerate, there are two Cases D1 and D2 (see Section 1.4).

Let $\zeta(y_3) \in C^\infty((-\infty, 0])$ satisfy $0 \le \zeta \le 1$, $\zeta(y_3) = 1$ for $-1/2 \le y_3 \le 0$ and 0 for $y_3 \le -1$.

Simple or Semisimple case Let $\begin{bmatrix} \mathbf{a}_\alpha \\ \mathbf{l}_\alpha \end{bmatrix}$ ($\alpha = 1, 2, 3$) be linearly independent eigenvectors of the eigenvalue problem (1.29) associated with the eigenvalues p_α ($\alpha = 1, 2, 3$, $\mathrm{Im}\, p_\alpha > 0$). Then Theorem 1.16 implies that their displacement parts $\mathbf{a}_\alpha$ ($\alpha = 1, 2, 3$) are linearly independent. Hence, for $\mathbf{v}$ in (2.45) there exist c_α ($\alpha = 1, 2, 3$) $\in \mathbb{C}$ such that

$$\mathbf{v} = \sum_{\alpha=1}^{3} c_\alpha \mathbf{a}_\alpha. \tag{2.47}$$

Taking account of (1.45), we set

$$\mathbf{F}_N(\mathbf{x}) = \eta\left(\sqrt{N} y'\right) \zeta\left(\sqrt{N} y_3\right) \sum_{\alpha=1}^{3} c_\alpha \mathbf{a}_\alpha e^{-\sqrt{-1}(y_1 + p_\alpha y_3)N}, \tag{2.48}$$

which is an extension of $f_N(y')$ and is meant to be an approximate solution to (2.35) and (2.36) with $f = f_N$ in a neighborhood of $\mathbf{x} = 0$ as $N \longrightarrow +\infty$. From (2.39) it follows that

$$< \Lambda(f_N), \overline{f_N} > = \int_\Omega \sum_{i,j,k,l=1}^{3} C_{ijkl}(\mathbf{x}) \frac{\partial(\mathbf{u}_N)_k}{\partial x_l} \frac{\partial\left(\overline{\mathbf{F}_N}\right)_i}{\partial x_j} d\mathbf{x}, \tag{2.49}$$

where $\mathbf{u}_N \in H^1(\Omega)$ is the solution to

$$\sum_{j,k,l=1}^{3} \frac{\partial}{\partial x_j}\left(C_{ijkl}(\mathbf{x})\frac{\partial(\mathbf{u}_N)_k}{\partial x_l}\right) = 0, \quad i = 1, 2, 3 \quad \text{in} \quad \Omega, \qquad \mathbf{u}_N|_{\partial\Omega} = f_N. \tag{2.50}$$

Let

$$\mathbf{u}_N = \mathbf{F}_N + \mathbf{r}_N. \tag{2.51}$$

Then $\mathbf{r}_N$ is the solution to

$$\sum_{j,k,l=1}^{3} \frac{\partial}{\partial x_j}\left(C_{ijkl}(\mathbf{x})\frac{\partial(\mathbf{r}_N)_k}{\partial x_l}\right) = -\sum_{j,k,l=1}^{3} \frac{\partial}{\partial x_j}\left(C_{ijkl}(\mathbf{x})\frac{\partial(\mathbf{F}_N)_k}{\partial x_l}\right),$$

$$i = 1, 2, 3 \quad \text{in} \quad \Omega,$$

$$\mathbf{r}_N|_{\partial\Omega} = 0. \tag{2.52}$$

 Springer

Using (2.51), we recast the right hand side of (2.49) as

$$\int_\Omega \sum_{i,j,k,l=1}^{3} C_{ijkl}(\mathbf{x}) \frac{\partial (\mathbf{F}_N)_k}{\partial x_l} \frac{\partial \left(\overline{\mathbf{F}_N}\right)_i}{\partial x_j}\, d\mathbf{x} + \int_\Omega \sum_{i,j,k,l=1}^{3} C_{ijkl}(\mathbf{x}) \frac{\partial (\mathbf{r}_N)_k}{\partial x_l} \frac{\partial \left(\overline{\mathbf{F}_N}\right)_i}{\partial x_j}\, d\mathbf{x}$$

and write

$$< \Lambda(\mathbf{f}_N), \overline{\mathbf{f}_N} > = I + II, \tag{2.53}$$

where

$$I = \int_\Omega \sum_{i,j,k,l=1}^{3} C_{ijkl}(\mathbf{x}) \frac{\partial (\mathbf{F}_N)_k}{\partial x_l} \frac{\partial \left(\overline{\mathbf{F}_N}\right)_i}{\partial x_j}\, d\mathbf{x},$$

$$II = \int_\Omega \sum_{i,j,k,l=1}^{3} C_{ijkl}(\mathbf{x}) \frac{\partial (\mathbf{r}_N)_k}{\partial x_l} \frac{\partial \left(\overline{\mathbf{F}_N}\right)_i}{\partial x_j}\, d\mathbf{x}.$$

It is sufficient to show that

$$\lim_{N\to\infty} I = (\mathbf{Z}_0\, \mathbf{v}, \mathbf{v})_{\mathbb{C}^3} \tag{2.54}$$

and

$$\lim_{N\to\infty} II = 0. \tag{2.55}$$

First we note from (2.40), (2.42) and (2.48) that

$$\frac{\partial \mathbf{F}_N}{\partial x_l} = -\sqrt{-1}\, N\, \eta\left(\sqrt{N}y'\right) \zeta\left(\sqrt{N}y_3\right)$$

$$\times \sum_{\alpha=1}^{3} c_\alpha\, \mathbf{a}_\alpha \left(m_l + p_\alpha(n_l - \phi_{y_1}m_l - \phi_{y_2}\ell_l)\right) e^{-\sqrt{-1}(y_1 + p_\alpha y_3)N}$$

$$+ \sqrt{N}\left(\left(\eta_{y_1}\left(\sqrt{N}y'\right)m_l + \eta_{y_2}\left(\sqrt{N}y'\right)\ell_l\right) \zeta\left(\sqrt{N}y_3\right)\right.$$

$$\left. + \eta\left(\sqrt{N}y'\right)\zeta'\left(\sqrt{N}y_3\right)(n_l - \phi_{y_1}m_l - \phi_{y_2}\ell_l)\right)$$

$$\times \sum_{\alpha=1}^{3} c_\alpha\, \mathbf{a}_\alpha e^{-\sqrt{-1}(y_1 + p_\alpha y_3)N},$$

and we write

$$\frac{\partial \mathbf{F}_N}{\partial x_l} \sim -\sqrt{-1}\, N\, \eta\left(\sqrt{N}y'\right) \zeta\left(\sqrt{N}y_3\right)$$

$$\times \sum_{\alpha=1}^{3} c_\alpha\, \mathbf{a}_\alpha \left(m_l + p_\alpha(n_l - \phi_{y_1}m_l - \phi_{y_2}\ell_l)\right) e^{-\sqrt{-1}(y_1 + p_\alpha y_3)N}. \tag{2.56}$$

Here and hereafter we use the notation $\sim$ to indicate that we are retaining the dominant terms in N for large N.

 Springer

Now we observe that

$$\lim_{N\to\infty} I = \lim_{N\to\infty} I', \tag{2.57}$$

where

$$I' = \int_\Omega \sum_{i,j,k,l=1}^{3} C_{ijkl}(\mathbf{0})\, \frac{\partial (\mathbf{F}_N)_k}{\partial x_l}\, \frac{\partial \overline{(\mathbf{F}_N)}_i}{\partial x_j}\, d\mathbf{x}.$$

In fact, after the change of variables $\mathbf{x} \longrightarrow \mathbf{y}$, it follows from (2.56) that

$$I' - I \sim N^2 \int_{\substack{|y'| \le \frac{1}{\sqrt{N}} \\ -\frac{1}{\sqrt{N}} \le y_3 \le 0}} \sum_{i,j,k,l=1}^{3} \left(C_{ijkl}(\mathbf{y}) - C_{ijkl}(\mathbf{0})\right) \eta\left(\sqrt{N} y'\right)^2 \varsigma \left(\sqrt{N} y_3\right)^2$$

$$\times \sum_{\alpha,\beta=1}^{3} c_\alpha \overline{c_\beta}\, (\mathbf{a}_\alpha)_k \left(\overline{\mathbf{a}_\beta}\right)_i \left(m_l + p_\alpha(n_l - \phi_{y_1} m_l - \phi_{y_2}\ell_l)\right)$$

$$\times \left(m_j + \overline{p_\beta}(n_j - \phi_{y_1} m_j - \phi_{y_2}\ell_j)\right) e^{-\sqrt{-1}(p_\alpha - \overline{p_\beta})y_3 N}\, d\mathbf{y},$$

where we have used the fact that the Jacobian related to the change of variables $\mathbf{x} \longrightarrow \mathbf{y}$ equals 1. Then by introducing the scaling transformation

$$z_i = \sqrt{N}\, y_i \quad (i = 1, 2), \qquad z_3 = N\, y_3, \tag{2.58}$$

we get

$$I' - I \sim \int_{\substack{|z'| \le 1 \\ -\sqrt{N} \le z_3 \le 0}} \sum_{i,j,k,l=1}^{3} \left(C_{ijkl}\left(\frac{z'}{\sqrt{N}}, \frac{z_3}{N}\right) - C_{ijkl}(\mathbf{0})\right) \eta(z')^2 \varsigma \left(\frac{z_3}{\sqrt{N}}\right)^2$$

$$\times \sum_{\alpha,\beta=1}^{3} c_\alpha \overline{c_\beta}\, (\mathbf{a}_\alpha)_k (\overline{\mathbf{a}_\beta})_i \left(m_l + p_\alpha \left(n_l - \phi_{y_1}(z'/\sqrt{N})\, m_l - \phi_{y_2}(z'/\sqrt{N})\, \ell_l\right)\right)$$

$$\times \left(m_j + \overline{p_\beta} \left(n_j - \phi_{y_1}(z'/\sqrt{N})\, m_j - \phi_{y_2}(z'/\sqrt{N})\, \ell_j\right)\right) e^{-\sqrt{-1}(p_\alpha - \overline{p_\beta})z_3}\, d\mathbf{z},$$

where $\mathbf{z} = (z_1, z_2, z_3) = (z', z_3)$. Since

$$\mathrm{Im}\left(p_\alpha - \overline{p_\beta}\right) > 0, \qquad \alpha, \beta = 1, 2, 3,$$

Lebesgue's dominated convergence theorem and the continuity of $\mathbf{C}(\mathbf{x}) = (C_{ijkl}(\mathbf{x}))_{i,j,k,l=1,2,3}$ at $\mathbf{x} = \mathbf{0}$ imply that the preceding integral tends to zero as $N \longrightarrow \infty$. This proves (2.57). To arrive at (2.54), we will prove

$$\lim_{N\to\infty} I' = (\mathbf{Z}_0\, \mathbf{v}, \mathbf{v})_{\mathbb{C}^3}. \tag{2.59}$$

From the divergence theorem it follows that

$$I' = \int_{\partial\Omega} \sum_{i,j,k,l=1}^{3} C_{ijkl}(\mathbf{0}) \frac{\partial(\mathbf{F}_N)_k}{\partial x_l} \left(\overline{\mathbf{F}_N}\right)_i \nu_j(\mathbf{x})\, d\sigma$$

$$- \int_{\Omega} \sum_{i,j,k,l=1}^{3} C_{ijkl}(\mathbf{0}) \frac{\partial^2(\mathbf{F}_N)_k}{\partial x_j \partial x_l} \left(\overline{\mathbf{F}_N}\right)_i d\mathbf{x}.$$

We write

$$I' = I_1 - I_2, \tag{2.60}$$

where

$$I_1 = \int_{\partial\Omega} \sum_{i,j,k,l=1}^{3} C_{ijkl}(\mathbf{0}) \frac{\partial(\mathbf{F}_N)_k}{\partial x_l} \left(\overline{\mathbf{F}_N}\right)_i \nu_j(\mathbf{x})\, d\sigma, \tag{2.61}$$

$$I_2 = \int_{\Omega} \sum_{i,j,k,l=1}^{3} C_{ijkl}(\mathbf{0}) \frac{\partial^2(\mathbf{F}_N)_k}{\partial x_j \partial x_l} \left(\overline{\mathbf{F}_N}\right)_i d\mathbf{x}, \tag{2.62}$$

and $\nu_j(\mathbf{x})$ is the j-th component of the unit outward normal to $\partial\Omega$ at $\mathbf{x} \in \partial\Omega$.[26] Then from (2.43), (2.48), (2.56) and $\zeta(0) = 1$, it follows that

$$I_1 \sim -\sqrt{-1}\, N \sum_{i,j,k,l=1}^{3} C_{ijkl}(\mathbf{0}) \int_{|y'| \le \frac{1}{\sqrt{N}}} \eta\left(\sqrt{N}y'\right)^2$$

$$\times \sum_{\alpha,\beta=1}^{3} c_\alpha \overline{c_\beta}\, (\mathbf{a}_\alpha)_k (\overline{\mathbf{a}_\beta})_i \left(m_l + p_\alpha(n_l - \phi_{y_1} m_l - \phi_{y_2} \ell_l)\right) \nu_j(y')\, J(y')\, dy_1 dy_2,$$

where we have used the fact that the surface element $d\sigma$ can be written from (2.40), (2.42) and (2.43) as

$$d\sigma = J(y')\, dy_1 dy_2, \qquad J(y') = \sqrt{1 + \phi_{y_1}^{\,2} + \phi_{y_2}^{\,2}}.$$

From the scaling transformation (2.58) and regrouping of the summation, it follows that

$$I_1 \sim -\sqrt{-1} \sum_{\alpha,\beta=1}^{3} c_\alpha \overline{c_\beta} \sum_{i,j,k,l=1}^{3} C_{ijkl}(\mathbf{0})\, (\mathbf{a}_\alpha)_k (\overline{\mathbf{a}_\beta})_i$$

$$\times \int_{|z'| \le 1} \eta(z')^2 \left(m_l + p_\alpha \left(n_l - \phi_{y_1}(z'/\sqrt{N}) m_l - \phi_{y_2}(z'/\sqrt{N}) \ell_l\right)\right)$$

$$\times \nu_j(z'/\sqrt{N})\, J(z'/\sqrt{N})\, dz'.$$

---

[26]Intuitively, since $\mathbf{F}_N$ is an approximate solution to (2.35) near $\mathbf{x} = \mathbf{0}$, we can expect that I_2 will be smaller than I_1 for large N. Moreover, since $\left[\sum_{j,k,l=1}^{3} C_{ijkl}(\mathbf{0}) \frac{\partial(\mathbf{F}_N)_k}{\partial x_l} \nu_j\right]_{i\downarrow 1,2,3}$ is the corresponding approximate traction on $\partial\Omega$, we can expect that I_1 approximates the bilinear pairing of the Neumann and Dirichlet data.

Recalling (2.41) and (2.44), from Lebesgue's dominated convergence theorem we get

$$\lim_{N \to \infty} I_1 = -\sqrt{-1} \sum_{\alpha,\beta=1}^{3} c_\alpha \overline{c_\beta} \sum_{i,j,k,l=1}^{3} C_{ijkl}(\mathbf{0})\, (\mathbf{a}_\alpha)_k (\overline{\mathbf{a}_\beta})_i (m_l + p_\alpha n_l) n_j^{\,27}$$

$$= -\sqrt{-1} \sum_{\alpha,\beta=1}^{3} c_\alpha \overline{c_\beta}\, \big([\mathbf{R}^T + p_\alpha \mathbf{T}]\mathbf{a}_\alpha,\, \mathbf{a}_\beta\big)_{\mathbb{C}^3}$$

$$= -\sqrt{-1} \sum_{\alpha,\beta=1}^{3} c_\alpha \overline{c_\beta}\, \big(\mathbf{l}_\alpha,\, \mathbf{a}_\beta\big)_{\mathbb{C}^3},$$

where we have appealed to (1.20) and (1.25) with $\mathbf{C} = \big(C_{ijkl}\big)_{i,j,k,l=1,2,3} = \mathbf{C}(\mathbf{0})$. Hence from (1.75) and (2.47) we obtain

$$\lim_{N \to \infty} I_1 = \sum_{\alpha,\beta=1}^{3} c_\alpha \overline{c_\beta}\, \big(\mathbf{Z}_0\, \mathbf{a}_\alpha,\, \mathbf{a}_\beta\big)_{\mathbb{C}^3} = \big(\mathbf{Z}_0\, \mathbf{v},\, \mathbf{v}\big)_{\mathbb{C}^3}. \tag{2.63}$$

Next we proceed to prove

$$\lim_{N \to \infty} I_2 = 0. \tag{2.64}$$

Differentiating (2.56) with respect to x_j, we get

$$\frac{\partial^2 \mathbf{F}_N}{\partial x_j \partial x_l} \sim -N^2\, \eta\big(\sqrt{N} y'\big)\, \zeta\big(\sqrt{N} y_3\big)$$

$$\times \sum_{\alpha=1}^{3} c_\alpha\, \mathbf{a}_\alpha\, \big(m_j + p_\alpha(n_j - \phi_{y_1} m_j - \phi_{y_2}\ell_j)\big)$$

$$\times \big(m_l + p_\alpha(n_l - \phi_{y_1} m_l - \phi_{y_2}\ell_l)\big)\, e^{-\sqrt{-1}(y_1 + p_\alpha y_3)N}.$$

Then it follows from (2.62) that

$$I_2 \sim -N^2 \sum_{i,j,k,l=1}^{3} C_{ijkl}(\mathbf{0}) \int_{\substack{|y'| \le \frac{1}{\sqrt{N}} \\ -\frac{1}{\sqrt{N}} \le y_3 \le 0}} \eta\big(\sqrt{N} y'\big)^2\, \zeta\big(\sqrt{N} y_3\big)^2$$

$$\times \sum_{\alpha,\beta=1}^{3} c_\alpha \overline{c_\beta}\, (\mathbf{a}_\alpha)_k (\overline{\mathbf{a}_\beta})_i \big(m_j + p_\alpha(n_j - \phi_{y_1} m_j - \phi_{y_2}\ell_j)\big)$$

$$\times \big(m_l + p_\alpha(n_l - \phi_{y_1} m_l - \phi_{y_2}\ell_l)\big)\, e^{-\sqrt{-1}(p_\alpha - \overline{p_\beta})y_3 N}\, d\mathbf{y},$$

---

[27] Recall that n_j denotes the j-th component of the outward unit normal to $\partial\Omega$ at $\mathbf{x} = \mathbf{0} \in \partial\Omega$. Hence $\nu_j(\mathbf{x})$ tends to n_j as $\mathbf{x} \to \mathbf{0}$.

and from the scaling transformation (2.58) and regrouping the summation it follows that

$$
I_2 \sim - \sum_{\alpha,\beta=1}^{3} c_\alpha \overline{c_\beta} \sum_{i,j,k,l=1}^{3} C_{ijkl}(\mathbf{0}) (\mathbf{a}_\alpha)_k (\overline{\mathbf{a}_\beta})_i \int_{\substack{|z'|\leq 1 \\ -\sqrt{N}\leq z_3 \leq 0}} \eta(z')^2 \zeta \left(\frac{z_3}{\sqrt{N}} \right)^2
$$
$$
\times \left(m_j + p_\alpha \left(n_j - \phi_{y_1}(z'/\sqrt{N}) m_j - \phi_{y_2}(z'/\sqrt{N}) \ell_j \right) \right)
$$
$$
\times \left(m_l + p_\alpha \left(n_l - \phi_{y_1}(z'/\sqrt{N}) m_l - \phi_{y_2}(z'/\sqrt{N}) \ell_l \right) \right) e^{-\sqrt{-1}(p_\alpha - \overline{p_\beta}) z_3} \, d\mathbf{z}.
$$

Recalling (2.41), (2.44) and $\zeta(0) = 1$, from Lebesgue's dominated convergence theorem we get

$$
\lim_{N\to\infty} I_2 = - \sum_{\alpha,\beta=1}^{3} c_\alpha \overline{c_\beta} \sum_{i,j,k,l=1}^{3} C_{ijkl}(\mathbf{0}) (\mathbf{a}_\alpha)_k (\overline{\mathbf{a}_\beta})_i (m_j + p_\alpha n_j)(m_l + p_\alpha n_l)
$$
$$
\times \int_{-\infty}^{0} e^{-\sqrt{-1}(p_\alpha - \overline{p_\beta}) z_3} \, dz_3,
$$

and from (1.20)

$$
= - \sum_{\alpha,\beta=1}^{3} c_\alpha \overline{c_\beta} \left([\mathbf{Q} + p_\alpha(\mathbf{R} + \mathbf{R}^T) + p_\alpha^2 \mathbf{T}] \mathbf{a}_\alpha, \mathbf{a}_\beta \right)_{\mathbb{C}^3} \int_{-\infty}^{0} e^{-\sqrt{-1}(p_\alpha - \overline{p_\beta}) z_3} \, dz_3,
$$

which, by (1.21), is equal to zero. This proves (2.64). Therefore, from (2.57), (2.60) and (2.63), we obtain (2.54).

Formula (2.55) can be proved essentially in a parallel way to [15, 63] (see also [55]). We give an outline of the proof: Since $\mathbf{F}_N$ is an approximate solution to (2.50), i.e., $\mathbf{F}_N$ is the dominant term of the solution $\mathbf{u}_N$ for large N, $\mathbf{r}_N$ can be taken as the remainder term. Then it can be expected that the integral II is of less order in N than the integral I for large N. In fact, applying the Lax-Milgram theorem to (2.52), we can estimate $\|\mathbf{r}_N\|_{H_0^1(\Omega)}$ from above by the $H^{-1}(\Omega)$ norm of the right hand side of (2.52). Then the integral II can be estimated to get (2.55). For the details we refer to the references above.

Degenerate Case D1 Let $\begin{bmatrix} \mathbf{a}_\alpha \\ \mathbf{l}_\alpha \end{bmatrix}$ ($\alpha = 1, 2$) be linearly independent eigenvectors of the eigenvalue problem (1.29) associated with the eigenvalues p_α ($\alpha = 1, 2$, $\mathrm{Im}\, p_\alpha > 0$) and $\begin{bmatrix} \mathbf{a}_3 \\ \mathbf{l}_3 \end{bmatrix}$ be a generalized eigenvector satisfying

$$
\mathbf{N} \begin{bmatrix} \mathbf{a}_3 \\ \mathbf{l}_3 \end{bmatrix} - p_2 \begin{bmatrix} \mathbf{a}_3 \\ \mathbf{l}_3 \end{bmatrix} = \begin{bmatrix} \mathbf{a}_2 \\ \mathbf{l}_2 \end{bmatrix}.
$$

Then Theorem 1.16 implies that their displacement parts $\mathbf{a}_\alpha$ ($\alpha = 1, 2, 3$) are linearly independent. Hence, for $\mathbf{v}$ in (2.45) there exist c_α ($\alpha = 1, 2, 3$) $\in \mathbb{C}$ such that

$$
\mathbf{v} = \sum_{\alpha=1}^{3} c_\alpha \mathbf{a}_\alpha. \tag{2.65}
$$

Taking account of (1.48), we set

$$\mathbf{F}_N(\mathbf{x}) = \eta(\sqrt{N}y')\,\zeta(\sqrt{N}y_3)\left(\sum_{\alpha=1}^{3} c_\alpha\,\mathbf{a}_\alpha e^{-\sqrt{-1}(y_1+p_\alpha y_3)N}\right.$$

$$\left. -\sqrt{-1}\,c_3\,y_3\,N\,\mathbf{a}_2 e^{-\sqrt{-1}(y_1+p_2 y_3)N}\right) \qquad (p_2=p_3). \tag{2.66}$$

Then $\mathbf{F}_N(\mathbf{x})$ is an extension of $\boldsymbol{f}_N(y')$. Hence equations (2.49) to (2.53) also apply in this case and we will prove (2.54).[28] From (2.40), (2.42) and (2.66) we get

$$\frac{\partial \mathbf{F}_N}{\partial x_l} \sim -\sqrt{-1}\,N\,\eta(\sqrt{N}y')\,\zeta(\sqrt{N}y_3) \tag{2.67}$$

$$\times\left[\sum_{\alpha=1}^{3} c_\alpha\,\mathbf{a}_\alpha\left(m_l + p_\alpha(n_l - \phi_{y_1}m_l - \phi_{y_2}\ell_l)\right)e^{-\sqrt{-1}(y_1+p_\alpha y_3)N}\right.$$

$$+ c_3\,\mathbf{a}_2\left(n_l - \phi_{y_1}m_l - \phi_{y_2}\ell_l - \sqrt{-1}\left(m_l + p_2(n_l - \phi_{y_1}m_l - \phi_{y_2}\ell_l)\right)Ny_3\right)$$

$$\left.\times\, e^{-\sqrt{-1}(y_1+p_2 y_3)N}\right].$$

We note that under the scaling transformation (2.58) the right hand side is of first order in N. Hence (2.57) also applies and we will prove (2.59). We divide I' into the two integrals I_1 and I_2 as in (2.60), (2.61), (2.62) and will prove (2.63) and (2.64). Since it follows from (2.43), (2.66) and (2.67) that

$$\mathbf{F}_N|_{\partial\Omega} = \eta\left(\sqrt{N}y'\right)\sum_{\alpha=1}^{3} c_\alpha\,\mathbf{a}_\alpha\,e^{-\sqrt{-1}\,y_1 N}$$

and

$$\frac{\partial \mathbf{F}_N}{\partial x_l}\bigg|_{\partial\Omega} \sim -\sqrt{-1}\,N\,\eta(\sqrt{N}y')\left[\sum_{\alpha=1}^{3} c_\alpha\,\mathbf{a}_\alpha\left(m_l + p_\alpha(n_l - \phi_{y_1}m_l - \phi_{y_2}\ell_l)\right)\right.$$

$$\left. +\, c_3\,\mathbf{a}_2\left(n_l - \phi_{y_1}m_l - \phi_{y_2}\ell_l\right)\right]e^{-\sqrt{-1}\,y_1 N},$$

---

[28] $\mathbf{F}_N(\mathbf{x})$ is meant to be an approximate solution to (2.50) in a neighborhood of $\mathbf{x}=0$ as $N\longrightarrow +\infty$.

we get

$$I_1 \sim -\sqrt{-1}\, N \sum_{i,j,k,l=1}^{3} C_{ijkl}(\mathbf{0}) \int_{|y'|\leq \frac{1}{\sqrt{N}}} \eta(\sqrt{N} y')^2$$

$$\times \left[\sum_{\alpha,\beta=1}^{3} c_\alpha \overline{c_\beta}\, (\mathbf{a}_\alpha)_k (\overline{\mathbf{a}_\beta})_i \left(m_l + p_\alpha (n_l - \phi_{y_1} m_l - \phi_{y_2} \ell_l) \right) \right.$$

$$\left. + c_3 \sum_{\beta=1}^{3} \overline{c_\beta}\, (\mathbf{a}_2)_k (\overline{\mathbf{a}_\beta})_i (n_l - \phi_{y_1} m_l - \phi_{y_2} \ell_l) \right] v_j(y')\, J(y')\, dy_1 dy_2.$$

From the scaling transformation (2.58) and regrouping of the summation, it follows that

$$I_1 \sim -\sqrt{-1} \sum_{\alpha,\beta=1}^{3} c_\alpha \overline{c_\beta} \sum_{i,j,k,l=1}^{3} C_{ijkl}(\mathbf{0})\, (\mathbf{a}_\alpha)_k (\overline{\mathbf{a}_\beta})_i$$

$$\times \int_{|z'|\leq 1} \eta(z')^2 \left(m_l + p_\alpha \left(n_l - \phi_{y_1}(z'/\sqrt{N})\, m_l - \phi_{y_2}(z'/\sqrt{N})\, \ell_l \right) \right)$$

$$\times v_j(z'/\sqrt{N})\, J(z'/\sqrt{N})\, dz'$$

$$-\sqrt{-1}\, c_3 \sum_{\beta=1}^{3} \overline{c_\beta} \sum_{i,j,k,l=1}^{3} C_{ijkl}(\mathbf{0})\, (\mathbf{a}_2)_k (\overline{\mathbf{a}_\beta})_i$$

$$\times \int_{|z'|\leq 1} \eta(z')^2 \left(n_l - \phi_{y_1}(z'/\sqrt{N})\, m_l - \phi_{y_2}(z'/\sqrt{N})\, \ell_l \right)$$

$$\times v_j(z'/\sqrt{N})\, J(z'/\sqrt{N})\, dz'.$$

Then Lebesgue's dominated convergence theorem implies that

$$\lim_{N\to\infty} I_1 = -\sqrt{-1} \sum_{\alpha,\beta=1}^{3} c_\alpha \overline{c_\beta} \sum_{i,j,k,l=1}^{3} C_{ijkl}(\mathbf{0})\, (\mathbf{a}_\alpha)_k (\overline{\mathbf{a}_\beta})_i (m_l + p_\alpha n_l) n_j$$

$$-\sqrt{-1}\, c_3 \sum_{\beta=1}^{3} \overline{c_\beta} \sum_{i,j,k,l=1}^{3} C_{ijkl}(\mathbf{0})\, (\mathbf{a}_2)_k (\overline{\mathbf{a}_\beta})_i\, n_l\, n_j$$

$$= -\sqrt{-1} \sum_{\alpha,\beta=1}^{3} c_\alpha \overline{c_\beta}\, \left([\mathbf{R}^T + p_\alpha \mathbf{T}] \mathbf{a}_\alpha, \mathbf{a}_\beta \right)_{\mathbb{C}^3}$$

$$-\sqrt{-1}\, c_3 \sum_{\beta=1}^{3} \overline{c_\beta}\, \left(\mathbf{T}\, \mathbf{a}_2, \mathbf{a}_\beta \right)_{\mathbb{C}^3}$$

$$= -\sqrt{-1} \sum_{\alpha,\beta=1}^{3} c_\alpha \overline{c_\beta}\, \left(\mathbf{l}_\alpha, \mathbf{a}_\beta \right)_{\mathbb{C}^3},$$

where we have appealed to (1.20), (1.25) and (1.50) with $\mathbf{C} = \left(C_{ijkl}\right)_{i,j,k,l=1,2,3} = \mathbf{C}(\mathbf{0})$. Hence from (1.75) and (2.65) we obtain

$$\lim_{N\to\infty} I_1 = \sum_{\alpha,\beta=1}^{3} c_\alpha \overline{c_\beta} \ (\mathbf{Z_0}\, \mathbf{a}_\alpha, \mathbf{a}_\beta)_{\mathbb{C}^3} = (\mathbf{Z_0}\, \mathbf{v}, \mathbf{v})_{\mathbb{C}^3}.$$

Next we prove (2.64). From (2.66) we get

$$\frac{\partial^2 \mathbf{F}_N}{\partial x_j \partial x_l} \sim - N^2 \, \eta(\sqrt{N} y') \, \varsigma\left(\sqrt{N} y_3\right)$$

$$\times \left[\sum_{\alpha=1}^{3} c_\alpha\, \mathbf{a}_\alpha \left(m_j + p_\alpha(n_j - \phi_{y_1} m_j - \phi_{y_2}\ell_j)\right) \right.$$

$$\times \left(m_l + p_\alpha(n_l - \phi_{y_1} m_l - \phi_{y_2}\ell_l)\right) e^{-\sqrt{-1}(y_1 + p_\alpha y_3)N}$$

$$+ c_3\, \mathbf{a}_2 \Big((n_j - \phi_{y_1} m_j - \phi_{y_2}\ell_j)\left(m_l + p_2(n_l - \phi_{y_1} m_l - \phi_{y_2}\ell_l)\right)$$

$$+ (m_j + p_2(n_j - \phi_{y_1} m_j - \phi_{y_2}\ell_j))\,(n_l - \phi_{y_1} m_l - \phi_{y_2}\ell_l)$$

$$- \sqrt{-1}\, N y_3 \left(m_j + p_2(n_j - \phi_{y_1} m_j - \phi_{y_2}\ell_j)\right)$$

$$\times \left. \left(m_l + p_2(n_l - \phi_{y_1} m_l - \phi_{y_2}\ell_l)\right)\Big) e^{-\sqrt{-1}(y_1 + p_2 y_3)N} \right].$$

Then under the scaling transformation (2.58) it follows from (2.41) and $\varsigma(0) = 1$ that

$$\sum_{j,k,l=1}^{3} C_{ijkl}(\mathbf{0})\frac{\partial^2 (\mathbf{F}_N)_k}{\partial x_j \partial x_l} \sim -N^2\, \eta(z')$$

$$\times \left[\sum_{\alpha=1}^{3} c_\alpha \sum_{j,k,l=1}^{3} C_{ijkl}(\mathbf{0})\, (\mathbf{a}_\alpha)_k (m_j + p_\alpha n_j)(m_l + p_\alpha n_l) e^{-\sqrt{-1}(y_1 + p_\alpha y_3)N} \right.$$

$$+ c_3 \sum_{j,k,l=1}^{3} C_{ijkl}(\mathbf{0})\, (\mathbf{a}_2)_k \Big(n_j(m_l + p_2 n_l) + (m_j + p_2 n_j)n_l$$

$$- \sqrt{-1}\, z_3(m_j + p_2 n_j)(m_l + p_2 n_l)\Big) e^{-\sqrt{-1}(y_1 + p_2 y_3)N} \Bigg]$$

$$= -N^2\, \eta(z') \left(\sum_{\alpha=1}^{3} c_\alpha \left[\mathbf{Q} + p_\alpha(\mathbf{R} + \mathbf{R}^T) + p_\alpha^2 \mathbf{T}\right] \mathbf{a}_\alpha e^{-\sqrt{-1}(y_1 + p_\alpha y_3)N} \right.$$

$$+ c_3 \left(\mathbf{R} + \mathbf{R}^T + 2 p_2 \mathbf{T} - \sqrt{-1}\, z_3\left[\mathbf{Q} + p_2\,(\mathbf{R} + \mathbf{R}^T) + p_2^2 \mathbf{T}\right]\right)$$

$$\times \left. \mathbf{a}_2\, e^{-\sqrt{-1}(y_1 + p_2 y_3)N}\right),$$

where we have used (1.20). From (1.21) for $\mathbf{a} = \mathbf{a}_1, \mathbf{a}_2$ and $p_2 = p_3$, we have

$$\sum_{j,k,l=1}^{3} C_{ijkl}(\mathbf{0}) \frac{\partial^2 (\mathbf{F}_N)_k}{\partial x_j \partial x_l} \sim -N^2 \eta(z') c_3$$

$$\times \left(\left[\mathbf{Q} + p_2(\mathbf{R} + \mathbf{R}^T) + p_2^2 \mathbf{T} \right] \mathbf{a}_3 + \left[\mathbf{R} + \mathbf{R}^T + 2p_2 \mathbf{T} \right] \mathbf{a}_2 \right) e^{-\sqrt{-1}(y_1 + p_2 y_3) N},$$

which is equal to zero by (1.52). We apply this to the integrand of I_2 as in the simple or semisimple case, which leads to

$$\lim_{N \to \infty} I_2 = 0.$$

Formula (2.55) can also be proved in the same way as in the simple or semisimple case (see the last paragraph of this case on page 83).

The degenerate case D2 remains to be considered. In this case, let $\begin{bmatrix} \mathbf{a}_\alpha \\ \mathbf{l}_\alpha \end{bmatrix}$ ($\alpha = 1, 2, 3$) be linearly independent eigenvector and generalized eigenvectors in Lemma 1.7. Then taking account of (1.55), we set

$$\mathbf{F}_N(\mathbf{x}) = \eta\left(\sqrt{N}y'\right) \zeta\left(\sqrt{N}y_3\right)$$

$$\times \left(\sum_{\alpha=1}^{3} c_\alpha \, \mathbf{a}_\alpha - \sqrt{-1}\, y_3 N (c_2 \mathbf{a}_1 + c_3 \mathbf{a}_2) - \frac{1}{2}(y_3 N)^2 c_3 \mathbf{a}_1 \right) e^{-\sqrt{-1}(y_1 + p_1 y_3) N}$$

$$(2.68)$$

as an extension of $\boldsymbol{f}_N(y')$, where c_α ($\alpha = 1, 2, 3$) $\in \mathbb{C}$ are chosen so that $\mathbf{v} = \sum_{\alpha=1}^{3} c_\alpha \mathbf{a}_\alpha$. The proof is parallel to the previous cases, where we need the formulas obtained in the proof of Lemma 1.7. We leave the details as Exercise 2-4. □

2.3.2.2 Reconstruction of Elasticity Tensor from Surface Impedance Tensor
By Theorem 2.16, we can reconstruct all the components of the surface impedance tensor $\mathbf{Z}_0$ at $\mathbf{x} = \mathbf{0} \in \partial\Omega$ from the localized Dirichlet to Neumann map Λ. The explicit formulas of $\mathbf{Z}_0$ are obtained for isotropic and transversely isotropic elastic materials in Section 1.8. Here we shall restrict our attention to the problem of reconstructing the components of $\mathbf{C}(\mathbf{0})$ from $\mathbf{Z}_0$ for these materials. In the process of reconstruction we choose some different unit tangents $\mathbf{m}$ in (2.45).

Theorem 2.17 *When the elastic material is isotropic, all the components of* $\mathbf{C}(\mathbf{0})$*, i.e., the two independent parameters* λ *and* μ *at* $\mathbf{x} = \mathbf{0}$*, can be reconstructed algebraically from* $\mathbf{Z}_0$*.*

Proof We take $\mathbf{m}$ so that

$$\mathbf{m} \times \mathbf{n} = \boldsymbol{\ell} = (\ell_1, \ell_2, 0),$$

Then from Theorem 1.24 it follows that

$$Z_{33} = \frac{2\mu(\lambda + 2\mu)}{\lambda + 3\mu}$$

Springer

and

$$\text{Im } Z_{13} = \frac{2\mu^2}{\lambda + 3\mu} \ell_2, \qquad \text{Im } Z_{23} = \frac{-2\mu^2}{\lambda + 3\mu} \ell_1.$$

Since it never holds that $\ell_1 = \ell_2 = 0$ under $\ell_3 = 0$, the quantity

$$a_1 = \frac{2\mu^2}{\lambda + 3\mu}$$

is recovered from $\text{Im } Z_{13}$ or $\text{Im } Z_{23}$. Since

$$\frac{Z_{33}}{a_1} = \frac{(\lambda + 2\mu)}{\mu} = \frac{\lambda}{\mu} + 2,$$

the quantity

$$a_2 = \frac{\lambda}{\mu}$$

is recovered from Z_{33} and a_1. Since

$$Z_{33} = 2\mu \frac{a_2 + 2}{a_2 + 3},$$

μ is reconstructed from Z_{33} and a_2. Then

$$\lambda = \mu\, a_2$$

is reconstructed. $\qquad\square$

Theorem 2.18 *Let the elastic material in question be transversely isotropic. Suppose that the normal to $\partial\Omega$ at $\mathbf{x} = \mathbf{0}$ does not coincide with the direction of the axis of rotational symmetry of the material at $\mathbf{x} = \mathbf{0}$. Then all the components of $\mathbf{C}(\mathbf{0})$ (5 independent components) can be reconstructed algebraically from $\mathbf{Z}_0$.*

Proof (First Step) We shall recover the following four quantities:

$$a_1 = \sqrt{\frac{L(A - N)}{2}}, \qquad a_2 = \frac{\sqrt{AL}}{L + \sqrt{AC}}\sqrt{\left(\sqrt{AC} - F\right)\left(\sqrt{AC} + F + 2L\right)},$$

$$a_3 = \sqrt{\frac{C}{A}}, \qquad a_4 = \frac{L\left(F - \sqrt{AC}\right)}{L + \sqrt{AC}}. \tag{2.69}$$

We take $\mathbf{m}$ so that

$$\mathbf{m} \times \mathbf{n} = \boldsymbol{\ell} = (\ell_1, \ell_2, 0).$$

Then from Theorem 1.26, taking account that $\ell_1^2 + \ell_2^2 = 1$, we have

$$Z_{11} = a_1 \ell_1^2 + a_2 \ell_2^2, \qquad Z_{22} = a_1 \ell_2^2 + a_2 \ell_1^2, \qquad Z_{33} = a_2\, a_3,$$

$$Z_{12} = (a_1 - a_2)\, \ell_1 \ell_2, \qquad Z_{13} = -\sqrt{-1}\, a_4 \ell_2, \qquad Z_{23} = \sqrt{-1}\, a_4 \ell_1.$$

We easily see that a_1 and a_2 can be recovered from Z_{11} and Z_{22} when $\ell_1^2 \neq \ell_2^2$ and from Z_{11} and Z_{12} when $\ell_1^2 = \ell_2^2 \; (= 1/2)$. Furthermore, since the strong convexity condition (1.101) implies that $a_2 \neq 0$, a_3 is recovered from Z_{33} and a_4 is recovered from Z_{13} or Z_{23}.

Recall that in Theorem 1.26 we have taken the axis of rotational symmetry to be the 3-axis. If the unit outward normal $\mathbf{n}$ to $\partial\Omega$ at $\mathbf{x} = \mathbf{0}$ coincides with $(0, 0, \pm 1)$, we necessarily have $\ell_3 = 0$, and from the above step only the four quantities a_i ($i = 1, 2, 3, 4$) can be determined from $\mathbf{Z_0}$. There are five independent components A, C, F, L, N to be reconstructed. Hence it is impossible to reconstruct these five independent components at such a point.

(*Second Step*) We reconstruct the parameter L of the elastic tensor. We take $\mathbf{m}$ so that ℓ_3 is not equal to zero but very close to zero. The formulae for Z_{13} and Z_{23} in Theorem 1.26 are rewritten as

$$Z_{13} = -b_1\left(\ell_1^2 + \ell_2^2\right)\ell_1\ell_3 - \sqrt{-1}\,b_2\ell_2, \qquad Z_{23} = -b_1\left(\ell_1^2 + \ell_2^2\right)\ell_2\ell_3 + \sqrt{-1}\,b_2\ell_1,$$

where

$$b_1 = \frac{L(F + L)}{D}, \qquad b_2 = L\left\{-1 + \frac{F+L}{D}K\left(\ell_1^2 + \ell_2^2\right)\right\}.$$

Since b_1 and b_2 are real and $\ell_3 \neq 0, \pm 1$, we see that b_1 and b_2 can be recovered from Z_{13} or Z_{23}. Moreover, it follows that

$$-L + b_1 K\left(\ell_1^2 + \ell_2^2\right) = b_2. \qquad (2.70)$$

On the other hand, we have

$$K^2 = \ell_3^2 + \frac{2L}{A - N}\left(\ell_1^2 + \ell_2^2\right) = \ell_3^2 + \frac{L^2}{a_1^2}\left(\ell_1^2 + \ell_2^2\right),$$

and combining this with (2.70) we obtain a quadratic equation for L

$$\left\{1 - \left(\frac{b_1}{a_1}\right)^2\left(\ell_1^2 + \ell_2^2\right)^3\right\}L^2 + 2b_2\,L + b_2^2 - b_1^2\left(\ell_1^2 + \ell_2^2\right)^2\ell_3^2 = 0, \qquad (2.71)$$

from which L can be reconstructed.

(*Third Step*) We proceed to reconstruct the other parameters A, C, F and N of the elastic tensor. From (2.69) we have

$$a_2 = \frac{\sqrt{AL}}{L + \sqrt{AC}}\sqrt{\left(\sqrt{AC} - F\right)\left(\sqrt{AC} + F + 2L\right)} = \sqrt{\frac{\sqrt{AC}}{a_3}}\sqrt{-a_4\left(2 + \frac{a_4}{L}\right)}.$$

Since a_2, a_3, a_4 and L have already been recovered in the previous steps, we can recover $\sqrt{AC}$ from the above. Using $a_3 = \sqrt{C/A}$ again, we reconstruct A and C. Finally F and N are reconstructed from a_4 and a_1 in (2.69). $\qquad\square$

Thus, by Theorem 2.16, Theorem 2.17 and Theorem 2.18, reconstruction of $\mathbf{C}(\mathbf{x_0})$, $\mathbf{x_0} \in \partial\Omega$ from the Dirichlet to Neumann map localized around $\mathbf{x_0}$ is completed for isotropic and transversely isotropic elastic materials.

2.4 Comments and References

The fundamental solution $\mathbf{G}(\mathbf{x})$ for general anisotropic elasticity has been studied by many authors and from various approaches, and many expressions of $\mathbf{G}(\mathbf{x})$ have been obtained. The attempt to write down the fundamental solution from the standpoint of the Stroh formalism goes back to [45] and can be found in a fairly unified manner in [40]. We emphasize that Theorem 2.2 and Theorem 2.13 can be applied to any class of elastic materials (and also of piezoelectric materials), regardless of whether the corresponding Stroh's eigenvalue problem is simple, semisimple or degenerate and irrespective of the length of the Jordan chain in question.

The word "degenerate" means that Stroh's eigenvalue problem has multiple eigenvalues, which also implies that the contour integral in the complex plane derived from formula (2.2) has multiple poles in its integrand. Thus degenerate cases need special consideration. Nevertheless, it is important to study degenerate cases because degeneracy can occur very often from additional symmetries that crystals have. The standpoint here is to give the formula of $\mathbf{G}(\mathbf{x})$ with a mathematically rigorous proof. Thus, from Theorem 2.2, we obtain by a consistent method for isotropic and transversely isotropic materials explicit forms of $\mathbf{G}(\mathbf{x})$ written in terms of their elasticity tensors. It should be pointed out that using Cauchy's theory of residues, Ting and Lee [84] gave the formula of $\mathbf{G}(\mathbf{x})$ in terms of the eigenvalues of Stroh's eigenvalue problem, which can also be applied to any class of elastic materials.

For transversely isotropic materials, several efforts have been made to obtain an explicit form of the fundamental solution (see the references in [49, 54, 84]). The formula in Corollary 2.4 is a result from [54] and is given in a unified expression which is generally valid, regardless of whether the eigenvalue problem (1.29) is degenerate or not.[29]

On the other hand, as we have stated in the note of Corollary 2.4, for $\mathbf{x} \neq 0$ such that $\widehat{x}_3 = \pm 1$, i.e., $\widehat{x}_1^2 + \widehat{x}_2^2 = 0$, we must take the limit $\widehat{x}_1^2 + \widehat{x}_2^2 \longrightarrow 0$ in (2.5) to obtain $\mathbf{G}^{\mathrm{Trans}}(\mathbf{x})$. Ting and Lee [84] derived the formula which is uniformly valid regardless of whether $\widehat{x}_1^2 + \widehat{x}_2^2 = 0$ or not. The components S_{11}^0 and S_{22}^0 in [84] are written as

$$
S_{11}^0 = \frac{2}{(A-N)K} + \frac{L\cos^2\alpha + C\sin^2\alpha}{ALGH}
$$
$$
- \frac{1}{G(H+GK+K^2)}\left(\frac{2(H+\cos^2\alpha)}{A-N} + \frac{2GH}{(A-N)K} + \frac{C\sin^2\alpha}{AL}\right),
$$

$$
S_{22}^0 = \frac{1}{AG} + \frac{1}{G(H+GK+K^2)}\left(\frac{2(H+\cos^2\alpha)}{A-N} + \frac{2GH}{(A-N)K} + \frac{C\sin^2\alpha}{AL}\right),
$$

and the other S_{ij}^0's are the same as those in (2.8). We see that S_{11}^0 and S_{22}^0 given in the expressions above are not indeterminant forms when $\sin^2\alpha = \widehat{x}_1^2 + \widehat{x}_2^2 \longrightarrow 0$.

Rudimentary accounts of the theory of piezoelectricity are given, for example, in [62, 75]. Barnett and Lothe [11] extended the Stroh six-dimensional formalism for elasticity to the eight-dimensional formalism for piezoelectricity. Lemma 2.8 and

---

[29] After the work in [54] we have noticed that $\frac{D}{D'} = \frac{1}{LGH}$ with D' given therein. Hence the formula in Corollary 2.4 is simpler than that in [54].

Lemma 2.9 are the results reported in [2]. These lemmas lead to a mathematically rigorous proof of Theorem 2.6, which is a fundamental theorem in the integral formalism in a higher dimension ([42, 43]). The Stroh formalism has subsequently been extended [4, 5, 25] to cover anisotropic media which are at once piezoelectric, piezomagnetic, and magnetoelectric.

It was Calderón [16] who first posed the inverse boundary value problem for the conductivity equation, which can be taken as a mathematical formulation of electrical impedance tomography. Let Ω be a bounded domain in $\mathbb{R}^3$. Given a strictly positive conductivity $\gamma(\mathbf{x})$ on Ω, we apply a voltage potential f on the boundary $\partial\Omega$ and solve the Dirichlet problem

$$\sum_{j=1}^{3} \frac{\partial}{\partial x_j}\left(\gamma(\mathbf{x})\frac{\partial u}{\partial x_j}\right) = 0, \quad \text{in } \Omega, \qquad u\big|_{\partial\Omega} = f,$$

where $u = u(\mathbf{x})$ is the electric voltage potential. The resulting current density distribution on $\partial\Omega$ is

$$\Lambda(f) = \sum_{j=1}^{3} \gamma(\mathbf{x})\frac{\partial u}{\partial x_j}\, n_j\bigg|_{\partial\Omega},$$

where $\mathbf{n} = (n_1, n_2, n_3)$ is the unit outward normal to $\partial\Omega$. Then $\Lambda(f)$ gives the current distributions on $\partial\Omega$ (the Neumann data) produced by the voltage distributions on $\partial\Omega$ (the Dirichlet data), and we call Λ the Dirichlet to Neumann map for the conductivity equation. The inverse boundary value problem in question is to determine the conductivity from observations of the voltage distributions and the resulting current distributions on $\partial\Omega$. Mathematically this problem can be formulated as follows: Determine $\gamma(\mathbf{x})$ from Λ. Together with the uniqueness and the stability problems associated with it, the problem of obtaining the formulas for reconstructing $\gamma(\mathbf{x})$ from Λ has been studied by many authors (for example, [15, 51, 55, 67]). For an overview of these results as well as other works in inverse boundary value problems, we refer to [85].

The inverse boundary value problem considered in Section 2.3 is the analog of Calderón's problem in elasticity. However, many new difficulties arise because in elasticity the equations of equilibrium constitute an elliptic system. For isotropic elasticity, the problem of determining and reconstructing λ and μ from Λ is studied in [1, 58–60].

For anisotropic elasticity, there are works [53, 57, 60, 61] which use the facts that the Dirichlet to Neumann map (2.38) is a pseudodifferential operator of order 1 and that its full symbol (i.e., all the terms in the asymptotic expansion of the symbol of the pseudodifferential operator) can be written in terms of the surface impedance tensor and its derivatives. In these works, the boundary and the elasticity tensor are assumed to be C^∞ at every point. For transversely isotropic materials, Nakamura, Tanuma and Uhlmann [57] proved that the elasticity tensor and all its derivatives at the boundary can be reconstructed from the full symbol of the Dirichlet to Neumann map. Here using the full symbol of the Dirichlet to Neumann map corresponds to making measurements of the tractions on the boundary pertaining to all possible displacements on that boundary. In [57, 61], a layer stripping algorithm was also developed for elastic media, which gives approximation for the elasticity tensor beneath the boundary surface.

In Section 2.3, we have considered the problem of reconstructing the elasticity tensor at the boundary from the localized Dirichlet to Neumann map for isotropic and transversely isotropic materials, which is part of the results in [56]. We have used explicit Dirichlet data (2.45), which are compactly supported in a neighborhood of a boundary point. The choice of these data is largely inspired by the results in Chapter 1. In fact, their extensions $\mathbf{F}_N \in H^1(\Omega)$ in (2.48) and (2.66) have almost the same forms as (1.45) and (1.48), respectively, except for the cutoff functions $\eta(\sqrt{N}\,y')$ and $\zeta(\sqrt{N}\,y_3)$. It should be noted that the reconstruction method in Subsection 2.3.2 is along the same line as the studies in [15, 63]. In [63], the Lamé coefficient μ and the residual stress in a highly simplified model of elasticity are reconstructed from the localized Dirichlet to Neumann map.

For a general anisotropic material, the surface impedance tensor is reconstructed from the Dirichlet to Neumann map by Theorem 2.16, but the surface impedance tensor is not written explicitly in terms of the elasticity tensor. In this case, we may appeal to the numerical approach. Namely, we compute the approximate surface impedance tensors from the estimated values of the elasticity tensor through the integral representation (1.76) and compare them with the surface impedance tensors reconstructed from the Dirichlet to Neumann map. We may repeat this procedure to update the elasticity tensor. Recall that the surface impedance tensor is a function of $\boldsymbol{\ell} = \mathbf{m} \times \mathbf{n}$. Although a surface impedance tensor, which is Hermitian, provides only 9 equations for the components of the elasticity tensor, we can use as many surface impedance tensors as we please by choosing different unit tangents $\mathbf{m}$ at a fixed point on the boundary.

The study on reconstruction of the elasticity tensor beneath the boundary surface from the localized Dirichlet to Neumann map is still in progress [56]. As a first step in this approach, reconstruction of the normal derivatives of the elasticity tensor at the boundary from the localized Dirichlet to Neumann map should be considered.

There are many other problems for which the Stroh formalism is useful. These are problems of cracks, dislocations, inclusions and holes in anisotropic elastic media, and anisotropic elastic wedges and composites. For a review of these problems we refer to [77, 80].

2.5 Exercises

2-1 Following the suggestions below, derive the formula for $\mathbf{G}^{\text{Iso}}(\mathbf{x})$ directly from the integral in (2.2).

(a) For $\mathbf{m} = \mathbf{m}(\phi)$, $\mathbf{n} = \mathbf{n}(\phi)$ given by (1.31), we put

$$\mathbf{m} \times \mathbf{n} = \boldsymbol{\ell}\left(= \mathbf{e}_1 \times \mathbf{e}_2 = \frac{\mathbf{x}}{|\mathbf{x}|}\right).$$

Show that

$$\mathbf{T}(\phi) = (\lambda + 2\mu)\mathbf{n} \otimes \mathbf{n} + \mu(\mathbf{m} \otimes \mathbf{m} + \boldsymbol{\ell} \otimes \boldsymbol{\ell}),$$

where the tensor product $\mathbf{m} \otimes \mathbf{n}$ can be expressed as a 3×3 matrix as

$$\mathbf{m} \otimes \mathbf{n} = \left(m_i\, n_j\right)_{i\downarrow j \to 1,2,3}.^{30}$$

(b) Write $\mathbf{T}(\phi)^{-1}$ in terms of the orthonormal basis $\{\mathbf{e}_1, \mathbf{e}_2, \boldsymbol{\ell}\}$ in $\mathbb{R}^3$.

(c) Use the equalities

$$\int_{-\pi}^{\pi} \cos^2\phi\, d\phi = \int_{-\pi}^{\pi} \sin^2\phi\, d\phi = \pi, \qquad \int_{-\pi}^{\pi} \cos\phi\, \sin\phi\, d\phi = 0,$$

and

$$\boldsymbol{\ell} \otimes \boldsymbol{\ell} = \frac{x_i\, x_j}{|\mathbf{x}|^2}$$

to derive the formula for $\mathbf{G}^{\mathrm{Iso}}(\mathbf{x})$.

2-2 Is it possible to prove Corollary 1.11 for Jordan chains of arbitrary length by using Lemma 2.8 ?

2-3 (Inverse boundary value problem for two-dimensional elasticity [53]) We consider the inverse boundary value problem for two-dimensional elasticity given in Exercise 1-12. If we change the ranges where the indices i, j, k, l run from $\{1, 2, 3\}$ to $\{1, 2\}$, then the arguments in Subsection 2.3.1 apply in a parallel way. Hence we can define the Dirichlet to Neumann map for two-dimensional elasticity. Let $\mathbf{n} = (n_1, n_2)$ be the unit outward normal to $\partial\Omega$ at $\mathbf{x} = \mathbf{0} \in \partial\Omega$ and $\mathbf{m} = (m_1, m_2)$ be a unit tangent there. We put

$$y = \mathbf{m} \cdot \mathbf{x} = m_1 x_1 + m_2 x_2.$$

Let $\eta(y)$ be a C^1 function with compact support in $\mathbb{R}$ which satisfies

$$0 \le \eta \le 1, \qquad \int_{\mathbb{R}} \eta(y)^2\, dy = 1, \qquad \mathrm{supp}\,\eta \subset \{|y| < 1\}.$$

For any vector $\mathbf{v}$ in $\mathbb{C}^2$, we define the Dirichlet data $\boldsymbol{f}_N$ on $\partial\Omega$ by

$$\boldsymbol{f}_N = \eta\!\left(\sqrt{N}y\right) \mathbf{v}\, e^{-\sqrt{-1}\, yN}\Big|_{\partial\Omega}, \tag{2.73}$$

where N is a natural number which is large enough.

(a) Show that

$$\lim_{N \to \infty} N^{-\frac{1}{2}} < \Lambda(\boldsymbol{f}_N), \overline{\boldsymbol{f}_N} > = (\mathbf{Z}_0\, \mathbf{v}, \mathbf{v})_{\mathbb{C}^2} = \sum_{i,j=1}^{2} Z_{ij}\, v_j\, \overline{v}_i,$$

---

[30] Use the following properties of the tensor product in this exercise. Let $\{\mathbf{v}_1, \mathbf{v}_2, \mathbf{v}_3\}$ be an orthonormal basis in $\mathbb{R}^3$. Then

$$\mathbf{v}_1 \otimes \mathbf{v}_1 + \mathbf{v}_2 \otimes \mathbf{v}_2 + \mathbf{v}_3 \otimes \mathbf{v}_3 = \mathbf{I} \text{ (identity matrix)}.$$

For real $a, b, c \neq 0$

$$\left(a\, \mathbf{v}_1 \otimes \mathbf{v}_1 + b\, \mathbf{v}_2 \otimes \mathbf{v}_2 + c\, \mathbf{v}_3 \otimes \mathbf{v}_3\right)^{-1} = \frac{1}{a}\, \mathbf{v}_1 \otimes \mathbf{v}_1 + \frac{1}{b}\, \mathbf{v}_2 \otimes \mathbf{v}_2 + \frac{1}{c}\, \mathbf{v}_3 \otimes \mathbf{v}_3. \tag{2.72}$$

where $\mathbf{Z_0} = (Z_{ij})_{i,j=1,2,3}$ is the surface impedance tensor obtained in Exercise 1-12 with (1.145) evaluated at $\mathbf{x} = \mathbf{0} \in \partial\Omega$.

(b) Would it be possible to reconstruct the components of (1.145) from $\mathbf{Z_0}$? (From this we shall see that the rotational invariance of the surface impedance tensor reduces the information that the Dirichlet to Neumann map pertaining to Dirichlet data (2.73) carries about the elasticity tensor.)

(c) How about the case of isotropic elasticity? (Two-dimensional isotropic elasticity is characterized by $A = C = \lambda + 2\mu$, $F = \lambda$, $L = \mu$)

2-4 Use (2.68) to prove (2.54) when the eigenvalue problem (1.29) is degenerate and the length of the Jordan chain is three.

3 Rayleigh Waves in the Stroh Formalism

3.1 The Stroh Formalism for Dynamic Elasticity

In this Chapter we study dynamic deformations of the elastic medium. In the Cartesian coordinates (x_1, x_2, x_3), the equations of motion with zero body force can be written, by adding the acceleration term to the right hand side of the equilibrium equations (1.8), as

$$\sum_{j=1}^{3} \frac{\partial}{\partial x_j} \sigma_{ij} = \rho \frac{\partial^2}{\partial t^2} u_i, \qquad i = 1, 2, 3, \tag{3.1}$$

where $\sigma = (\sigma_{ij})_{i, j=1,2,3}$ is the stress tensor, ρ is the uniform mass density, t is the time, and $\boldsymbol{u} = \boldsymbol{u}(\mathbf{x}, t) = (u_1, u_2, u_3)$ is the displacement at the place $\mathbf{x} = (x_1, x_2, x_3)$ at time t. By (1.5), the equations of motion (3.1) can be written in terms of the displacement $\boldsymbol{u}$ as

$$\sum_{j,k,l=1}^{3} \frac{\partial}{\partial x_j} \left(C_{ijkl} \frac{\partial u_k}{\partial x_l} \right) = \rho \frac{\partial^2}{\partial t^2} u_i, \qquad i = 1, 2, 3. \tag{3.2}$$

Here $\mathbf{C} = (C_{ijkl})_{i, j,k,l=1,2,3}$ is the elasticity tensor, which has the minor symmetries (1.4) and the major symmetry (1.6) and satisfies the strong convexity condition (1.7).

Throughout this chapter we assume that the elastic medium in question is homogeneous. Then the elasticity tensor $\mathbf{C} = (C_{ijkl})_{i, j,k,l=1,2,3}$ is independent of $\mathbf{x}$, so that (3.2) becomes

$$\sum_{j,k,l=1}^{3} C_{ijkl} \frac{\partial^2 u_k}{\partial x_j \partial x_l} = \rho \frac{\partial^2}{\partial t^2} u_i, \qquad i = 1, 2, 3. \tag{3.3}$$

Let $\mathbf{x} = (x_1, x_2, x_3)$ be the position vector and let $\mathbf{m} = (m_1, m_2, m_3), \mathbf{n} = (n_1, n_2, n_3)$ be orthogonal unit vectors in $\mathbb{R}^3$. We consider motions of a homogeneous elastic medium which occupies the half-space $\mathbf{n} \cdot \mathbf{x} = n_1 x_1 + n_2 x_2 + n_3 x_3 \leq 0$, and seek solutions to (3.3) of the form

$$\boldsymbol{u} = (u_1, u_2, u_3) = \mathbf{a} \, e^{-\sqrt{-1} \, k(\mathbf{m} \cdot \mathbf{x} + p \, \mathbf{n} \cdot \mathbf{x} - v \, t)} \in \mathbb{C}^3 \tag{3.4}$$

in $\mathbf{n} \cdot \mathbf{x} \leq 0$. When Im p, the imaginary part of $p \in \mathbb{C}$, is positive, a solution of the preceding form describes a surface wave in $\mathbf{n} \cdot \mathbf{x} \leq 0$; it propagates along the surface $\mathbf{n} \cdot \mathbf{x} = 0$ in the direction of $\mathbf{m}$ with the wave number k and the phase velocity $v > 0$, has the polarization defined by a constant vector $\mathbf{a}$, and decays exponentially as $\mathbf{n} \cdot \mathbf{x} \longrightarrow -\infty$.

Let us determine the equations that $\mathbf{a} \in \mathbb{C}^3$ and $p \in \mathbb{C}$ in (3.4) must satisfy. Substituting (3.4) into (3.3) and noting that

$$\frac{\partial \boldsymbol{u}}{\partial x_j} = -\sqrt{-1} \, k(m_j + p n_j) \, \mathbf{a} \, e^{-\sqrt{-1} \, k(\mathbf{m} \cdot \mathbf{x} + p \, \mathbf{n} \cdot \mathbf{x} - v \, t)}$$

and

$$\frac{\partial \boldsymbol{u}}{\partial t} = \sqrt{-1}\, k\, v\, \boldsymbol{a}\, e^{-\sqrt{-1}\, k(\mathbf{m}\cdot\mathbf{x} + p\,\mathbf{n}\cdot\mathbf{x} - v\,t)},$$

we get

$$\left(\sum_{j,l=1}^{3} C_{ijkl}\, (m_j + pn_j)(m_l + pn_l) - \rho\, v^2 \delta_{ik} \right)_{i\downarrow k \to 1,2,3} \boldsymbol{a}$$

$$= \left(\sum_{j,l=1}^{3} C_{ijkl}\, m_j m_l - \rho\, v^2 \delta_{ik} + p \left(\sum_{j,l=1}^{3} C_{ijkl}\, m_j n_l + \sum_{j,l=1}^{3} C_{ijkl}\, n_j m_l \right) \right.$$

$$\left. + p^2 \sum_{j,l=1}^{3} C_{ijkl}\, n_j n_l \right)_{i\downarrow k \to 1,2,3} \boldsymbol{a} = 0, \tag{3.5}$$

where δ_{ik} is Kronecker's delta symbol. Now we introduce the dynamic elastic coefficients

$$C_{ijkl}^{d} = C_{ijkl} - \rho\, v^2 m_j m_l \delta_{ik}, \qquad i, j, k, l = 1, 2, 3 \tag{3.6}$$

and the 3×3 real matrices

$$\mathbf{Q} = \left(\sum_{j,l=1}^{3} C_{ijkl}^{d}\, m_j m_l \right)_{i\downarrow k \to 1,2,3}, \qquad \mathbf{R} = \left(\sum_{j,l=1}^{3} C_{ijkl}^{d}\, m_j n_l \right)_{i\downarrow k \to 1,2,3},$$

$$\mathbf{T} = \left(\sum_{j,l=1}^{3} C_{ijkl}^{d}\, n_j n_l \right)_{i\downarrow k \to 1,2,3}. \tag{3.7}$$

Since $\mathbf{m}$ and $\mathbf{n}$ are orthonormal, we have

$$\sum_{j=1}^{3} m_j^2 = 1, \qquad \sum_{j=1}^{3} n_j^2 = 1, \qquad \sum_{j=1}^{3} m_j n_j = 0. \tag{3.8}$$

Then it follows from (3.6) that

$$\mathbf{Q} = \left(\sum_{j,l=1}^{3} C_{ijkl}\, m_j m_l - \rho\, v^2 \delta_{ik} \right)_{i\downarrow k \to 1,2,3}, \tag{3.9}$$

$$\mathbf{R} = \left(\sum_{j,l=1}^{3} C_{ijkl}\, m_j n_l \right)_{i\downarrow k \to 1,2,3}, \qquad \mathbf{T} = \left(\sum_{j,l=1}^{3} C_{ijkl}\, n_j n_l \right)_{i\downarrow k \to 1,2,3}.$$

Therefore (3.5) can be written as

$$[\mathbf{Q} + p(\mathbf{R} + \mathbf{R}^T) + p^2 \mathbf{T}]\boldsymbol{a}$$

$$= \left(\sum_{j,l=1}^{3} C_{ijkl}^{d}\, (m_j + pn_j)(m_l + pn_l) \right)_{i\downarrow k \to 1,2,3} \boldsymbol{a} = 0. \tag{3.10}$$

For the existence of $\mathbf{a} \neq \mathbf{0}$ that satisfies (3.10), we observe that p must be a solution to the sextic equation

$$\det[\mathbf{Q} + p(\mathbf{R} + \mathbf{R}^T) + p^2\mathbf{T}]$$

$$= \det\left(\sum_{j,l=1}^{3} C_{ijkl}^d (m_j + pn_j)(m_l + pn_l)\right)_{i\downarrow k \to 1,2,3} = 0. \qquad (3.11)$$

Next we give the traction $\mathbf{t}$ on the surface $\mathbf{n} \cdot \mathbf{x} = 0$ produced by the surface-wave solution (3.4). From (1.24), (3.4) and (3.9) it follows that

$$\mathbf{t} = \left(\sum_{j,k,l=1}^{3} C_{ijkl}\frac{\partial u_k}{\partial x_l} n_j\right)_{i\downarrow 1,2,3}\Bigg|_{\mathbf{n}\cdot\mathbf{x}=0}$$

$$= -\sqrt{-1}\,k\left(\sum_{j,l=1}^{3} C_{ijkl}(m_l + pn_l)\,n_j\right)_{i\downarrow k \to 1,2,3} \mathbf{a}\,e^{-\sqrt{-1}\,k(\mathbf{m}\cdot\mathbf{x}-v\,t)}$$

$$= -\sqrt{-1}\,k\,[\mathbf{R}^T + p\mathbf{T}]\,\mathbf{a}\,e^{-\sqrt{-1}\,k(\mathbf{m}\cdot\mathbf{x}-v\,t)}.$$

Hence we define a vector $\mathbf{l} \in \mathbb{C}^3$ as

$$\mathbf{l} = [\mathbf{R}^T + p\mathbf{T}]\,\mathbf{a}. \qquad (3.12)$$

Then

$$\mathbf{t} = -\sqrt{-1}\,k\mathbf{l}\,e^{-\sqrt{-1}\,k(\mathbf{m}\cdot\mathbf{x}-v\,t)} \qquad (3.13)$$

is the traction on the surface $\mathbf{n} \cdot \mathbf{x} = 0$ produced by (3.4).

Equations (3.10) and (3.12) have exactly the same form as (1.21) and (1.25) in static elasticity, respectively. Therefore, (3.10) and (3.12) can be recast in Stroh's six-dimensional eigenrelation

$$\mathbf{N}\begin{bmatrix} \mathbf{a} \\ \mathbf{l} \end{bmatrix} = p\begin{bmatrix} \mathbf{a} \\ \mathbf{l} \end{bmatrix}, \qquad (3.14)$$

where $\mathbf{N}$ is the 6×6 real matrix defined by

$$\mathbf{N} = \begin{bmatrix} -\mathbf{T}^{-1}\mathbf{R}^T & \mathbf{T}^{-1} \\ -\mathbf{Q} + \mathbf{R}\mathbf{T}^{-1}\mathbf{R}^T & -\mathbf{R}\mathbf{T}^{-1} \end{bmatrix} \qquad (3.15)$$

and $\mathbf{Q}, \mathbf{R}$ and $\mathbf{T}$ are defined by (3.7).

When $v = 0$, it follows that $C_{ijkl}^d = C_{ijkl}$ and the solutions p_α ($1 \leq \alpha \leq 6$) to (3.11), i.e., the eigenvalues of $\mathbf{N}$, are not real (Lemma 1.1). As v increases from $v = 0$, at some point equation (3.11) ceases to have only complex roots. Since we are concerned with the surface-wave solution (3.4), where $\text{Im}\,p > 0$, we shall restrict our attention to the range of v for which all the solutions to (3.11) are complex.

Let $\tilde{\mathbf{m}} = (\tilde{m}_1, \tilde{m}_2, \tilde{m}_3)$ and $\tilde{\mathbf{n}} = (\tilde{n}_1, \tilde{n}_2, \tilde{n}_3)$ be orthogonal unit vectors in $\mathbb{R}^3$ which are obtained by rotating the orthogonal unit vectors $\mathbf{m}$ and $\mathbf{n}$ around their vector product $\mathbf{m} \times \mathbf{n}$ by an angle ϕ ($-\pi \leq \phi < \pi$) so that

$$\tilde{\mathbf{m}} = \tilde{\mathbf{m}}(\phi) = \mathbf{m}\cos\phi + \mathbf{n}\sin\phi, \qquad \tilde{\mathbf{n}} = \tilde{\mathbf{n}}(\phi) = -\mathbf{m}\sin\phi + \mathbf{n}\cos\phi. \qquad (3.16)$$

Let $\mathbf{Q}(\phi)$, $\mathbf{R}(\phi)$ and $\mathbf{T}(\phi)$ be the 3×3 real matrices given by

$$\mathbf{Q}(\phi) = \left(\sum_{j,l=1}^{3} C_{ijkl}^d \, \tilde{m}_j \tilde{m}_l\right)_{i\downarrow k \rightarrow 1,2,3} \quad , \quad \mathbf{R}(\phi) = \left(\sum_{j,l=1}^{3} C_{ijkl}^d \, \tilde{m}_j \tilde{n}_l\right)_{i\downarrow k \rightarrow 1,2,3} \quad ,$$

$$\mathbf{T}(\phi) = \left(\sum_{j,l=1}^{3} C_{ijkl}^d \, \tilde{n}_j \tilde{n}_l\right)_{i\downarrow k \rightarrow 1,2,3} \quad , \tag{3.17}$$

where C_{ijkl}^d ($i, j, k, l = 1, 2, 3$) are the dynamic elastic coefficients in (3.6).[31] Then $\mathbf{Q}(0)$, $\mathbf{R}(0)$ and $\mathbf{T}(0)$ are equal to $\mathbf{Q}, \mathbf{R}$ and $\mathbf{T}$ in (3.7), respectively. We note that C_{ijkl}^d no longer have the minor symmetries because of the term $\rho v^2 m_j m_l \delta_{ik}$, but we see that they still enjoy the major symmetry

$$C_{ijkl}^d = C_{klij}^d, \qquad i, j, k, l = 1, 2, 3. \tag{3.18}$$

Hence it follows from (3.17) that $\mathbf{Q}(\phi)$ and $\mathbf{T}(\phi)$ are symmetric for all ϕ. We also note that $\mathbf{Q}(\phi)$, $\mathbf{R}(\phi)$ and $\mathbf{T}(\phi)$ are π-periodic in ϕ, because they are quadratic in $\cos\phi$ and $\sin\phi$ by (3.17). Moreover, it follows that

$$\mathbf{Q}\left(\phi + \frac{\pi}{2}\right) = \mathbf{T}(\phi), \quad \mathbf{R}\left(\phi + \frac{\pi}{2}\right) = -\mathbf{R}(\phi)^T, \quad \mathbf{T}\left(\phi + \frac{\pi}{2}\right) = \mathbf{Q}(\phi). \tag{3.19}$$

Definition 3.1 The limiting velocity $v_L = v_L(\mathbf{m}, \mathbf{n})$ is the lowest velocity for which the matrices $\mathbf{Q}(\phi)$ and $\mathbf{T}(\phi)$ become singular for some angle ϕ:

$$v_L = \inf\{v > 0 \mid \exists \phi; \, \det \mathbf{Q}(\phi) = 0\}$$

$$= \inf\{v > 0 \mid \exists \phi; \, \det \mathbf{T}(\phi) = 0\}.[32] \tag{3.20}$$

The interval $0 < v < v_L$ is called the subsonic range. It will be observed in the example after Lemma 3.3 and also in Section 3.4 that v_L for isotropic elasticity is given by

$$v_L^{\text{Iso}} = \sqrt{\frac{\mu}{\rho}}. \tag{3.21}$$

Lemma 3.2

(1) *The symmetric matrices $\mathbf{Q}(\phi)$ and $\mathbf{T}(\phi)$ are positive definite for all ϕ if and only if $0 \leq v < v_L$.*
(2) *The solutions p_α ($1 \leq \alpha \leq 6$) to the sextic equation (3.11), i.e., the eigenvalues of $\mathbf{N}$, are not real and they occur in complex conjugate pairs if and only if $0 \leq v < v_L$.*

---

[31] We do not rotate $\mathbf{m}$ in (3.6).

[32] The last equality follows from (3.19).

Proof To prove (1), we note from (3.6), (3.8), (3.16) and (3.17) that

$$\mathbf{Q}(\phi) = \left(\sum_{j,l=1}^{3}(C_{ijkl} - \rho\, v^2 m_j m_l \delta_{ik})\, \tilde{m}_j \tilde{m}_l\right)_{i\downarrow k \to 1,2,3}$$

$$= \left(\sum_{j,l=1}^{3} C_{ijkl}\, \tilde{m}_j \tilde{m}_l\right)_{i\downarrow k \to 1,2,3} - \rho\, v^2 \cos^2 \phi\, \mathbf{I}, \tag{3.22}$$

where $\mathbf{I}$ is the 3×3 identity matrix. By Lemma 1.1, the real symmetric matrix[33]

$$\left(\sum_{j,l=1}^{3} C_{ijkl}\, \tilde{m}_j \tilde{m}_l\right)_{i\downarrow k \to 1,2,3}$$

is positive definite. Hence it has three positive eigenvalues, which we write as

$$\lambda_i(\phi)\ (i = 1, 2, 3), \qquad 0 < \lambda_1(\phi) \le \lambda_2(\phi) \le \lambda_3(\phi). \tag{3.23}$$

Then by (3.22), the eigenvalues of $\mathbf{Q}(\phi)$ are

$$\lambda_i(\phi) - \rho\, v^2 \cos^2 \phi, \qquad i = 1, 2, 3. \tag{3.24}$$

These are strictly decreasing functions of v if $\cos\phi \ne 0$. Then, as v increases from zero, $\mathbf{Q}(\phi)$ first becomes singular when v is equal to

$$v_\phi = \sqrt{\frac{\lambda_1(\phi)}{\rho}}\, \frac{1}{|\cos\phi|}, \tag{3.25}$$

while $\mathbf{Q}(\phi)$ is positive definite for $0 \le v < v_\phi$. Since $\mathbf{Q}(\phi)$ is π-periodic in ϕ, we can restrict the range of ϕ to the interval $-\frac{\pi}{2} < \phi < \frac{\pi}{2}$. Thus, we put

$$v_L = \min_{-\frac{\pi}{2} < \phi < \frac{\pi}{2}} v_\phi. \tag{3.26}$$

By the first equality in (3.19), the assertion (1) of the lemma also holds for $\mathbf{T}(\phi)$.

Now we prove (2). Suppose that $0 \le v < v_L$. Then the same proof of (2) of Lemma 1.1 can be applied to prove that the solutions p_α are not real and that they occur in complex conjugate pairs if we replace C_{ijkl} there with C_{ijkl}^d in (3.6). Now suppose there exists some v such that $v_L \le v$ and such that the matrix $\left(\sum_{j,l=1}^{3} C_{ijkl}^d\right.$ $\left.(m_j + pn_j)(m_l + pn_l)\right)_{i\downarrow k \to 1,2,3}$ is invertible for all $p \in \mathbb{R}$. From the definition of v_L

---

[33] Later in Subsection 3.6.1, we shall call this matrix the acoustical tensor (see (3.130)).

and the monotonicity of the eigenvalues of $\mathbf{Q}(\phi)$ in v (see (3.24)), there exists $\hat{\phi}$ $\left(-\frac{\pi}{2} < \hat{\phi} < \frac{\pi}{2}\right)$ such that $\mathbf{Q}(\hat{\phi})$ is singular. By (3.16), and (3.17), $\mathbf{Q}(\hat{\phi})$ is written as

$$\mathbf{Q}(\hat{\phi}) = \left(\sum_{j,l=1}^{3} C_{ijkl}^{d}(m_j \cos\hat{\phi} + n_j \sin\hat{\phi})(m_l \cos\hat{\phi} + n_l \sin\hat{\phi}) \right)_{i\downarrow k \rightarrow 1,2,3}$$

$$= \cos^6\hat{\phi} \left(\sum_{j,l=1}^{3} C_{ijkl}^{d}(m_j + n_j \tan\hat{\phi})(m_l + n_l \tan\hat{\phi}) \right)_{i\downarrow k \rightarrow 1,2,3},$$

which leads to a contradiction when $p = \tan\hat{\phi}$. $\square$

Henceforth, as in Chapter 1 we take

$$\mathrm{Im}\, p_\alpha > 0, \qquad \alpha = 1, 2, 3.$$

3.2 Basic Theorems and Integral Formalism

On the basis of Lemma 3.2, we can see that the results in Section 1.3 to Section 1.7 apply in a parallel way to dynamic elasticity when $0 \leq v < v_L$. Hence in this section we summarize the fundamental results in the Stroh formalism for dynamic elasticity, which will be needed for the analysis of Rayleigh waves in the subsequent sections.

Let $\mathbf{m}$ and $\mathbf{n}$ be orthogonal unit vectors in $\mathbb{R}^3$. First we give the general forms of the surface-wave solutions to (3.3) in the half-space $\mathbf{n} \cdot \mathbf{x} \leq 0$ which decay exponentially as $\mathbf{n} \cdot \mathbf{x} \longrightarrow -\infty$ and have direction of propagation $\mathbf{m}$, wave number k and phase velocity v on the surface $\mathbf{n} \cdot \mathbf{x} = 0$. At the same time we give the corresponding tractions on the surface $\mathbf{n} \cdot \mathbf{x} = 0$ produced by the surface-wave solutions. These forms are different according to the degeneracy of Stroh's eigenvalue problem (3.14).

Lemma 3.3 *Let* $0 \leq v < v_L$.

Simple or Semisimple Case *Let* $\begin{bmatrix} \mathbf{a}_\alpha \\ \mathbf{l}_\alpha \end{bmatrix}$ *($\alpha = 1, 2, 3$) be linearly independent eigenvectors of the eigenvalue problem (3.14) pertaining to the eigenvalues p_α ($\alpha = 1, 2, 3$, $\mathrm{Im}\, p_\alpha > 0$), respectively. As we have seen from (3.4) and (3.13), the surface-wave solution is given by*

$$u = \sum_{\alpha=1}^{3} c_\alpha\, \mathbf{a}_\alpha\, e^{-\sqrt{-1}\,k(\mathbf{m}\cdot\mathbf{x}+p_\alpha\mathbf{n}\cdot\mathbf{x}-v t)} \tag{3.27}$$

and the corresponding traction on the surface $\mathbf{n} \cdot \mathbf{x} = 0$ produced by the solution above is given by

$$t = -\sqrt{-1}\,k \sum_{\alpha=1}^{3} c_\alpha\, \mathbf{l}_\alpha\, e^{-\sqrt{-1}\,k(\mathbf{m}\cdot\mathbf{x}-v t)}, \tag{3.28}$$

where c_α ($1 \leq \alpha \leq 3$) are arbitrary complex constants.

 Springer

Degenerate Case D1 *Let* p_1, $p_2 = p_3$ *be the eigenvalues of* $\mathbf{N}$ *with positive imaginary parts. Let* $\begin{bmatrix} \mathbf{a}_\alpha \\ \mathbf{l}_\alpha \end{bmatrix}$ $(\alpha = 1, 2)$ *be linearly independent eigenvectors pertaining to* p_α $(\alpha = 1, 2)$, *respectively, and let* $\begin{bmatrix} \mathbf{a}_3 \\ \mathbf{l}_3 \end{bmatrix}$ *be a generalized eigenvector satisfying*

$$\mathbf{N}\begin{bmatrix} \mathbf{a}_3 \\ \mathbf{l}_3 \end{bmatrix} - p_2 \begin{bmatrix} \mathbf{a}_3 \\ \mathbf{l}_3 \end{bmatrix} = \begin{bmatrix} \mathbf{a}_2 \\ \mathbf{l}_2 \end{bmatrix}.$$

Then the surface-wave solution is given by

$$\boldsymbol{u} = \sum_{\alpha=1}^{2} c_\alpha \, \mathbf{a}_\alpha \, e^{-\sqrt{-1}\,k(\mathbf{m}\cdot\mathbf{x}+p_\alpha\mathbf{n}\cdot\mathbf{x}-vt)}$$

$$+ c_3 \left(\mathbf{a}_3 - \sqrt{-1}\,k(\mathbf{n}\cdot\mathbf{x})\,\mathbf{a}_2 \right) e^{-\sqrt{-1}\,k(\mathbf{m}\cdot\mathbf{x}+p_2\mathbf{n}\cdot\mathbf{x}-vt)}, \tag{3.29}$$

and the corresponding traction on the surface $\mathbf{n} \cdot \mathbf{x} = 0$ *produced by the preceding solution is given by*

$$\boldsymbol{t} = -\sqrt{-1}\,k \sum_{\alpha=1}^{3} c_\alpha \, \mathbf{l}_\alpha \, e^{-\sqrt{-1}\,k(\mathbf{m}\cdot\mathbf{x}-vt)}, \tag{3.30}$$

where c_α $(1 \le \alpha \le 3)$ *are arbitrary complex constants.*

Degenerate Case D2 *Let* $p_1 = p_2 = p_3$ *be the triple eigenvalue of* $\mathbf{N}$ *with positive imaginary part and let* $\begin{bmatrix} \mathbf{a}_1 \\ \mathbf{l}_1 \end{bmatrix}$ *be a corresponding eigenvector. Let* $\begin{bmatrix} \mathbf{a}_\alpha \\ \mathbf{l}_\alpha \end{bmatrix}$ $(\alpha = 2, 3)$ *be generalized eigenvectors satisfying*

$$\mathbf{N}\begin{bmatrix} \mathbf{a}_2 \\ \mathbf{l}_2 \end{bmatrix} - p_1 \begin{bmatrix} \mathbf{a}_2 \\ \mathbf{l}_2 \end{bmatrix} = \begin{bmatrix} \mathbf{a}_1 \\ \mathbf{l}_1 \end{bmatrix}, \quad \mathbf{N}\begin{bmatrix} \mathbf{a}_3 \\ \mathbf{l}_3 \end{bmatrix} - p_1 \begin{bmatrix} \mathbf{a}_3 \\ \mathbf{l}_3 \end{bmatrix} = \begin{bmatrix} \mathbf{a}_2 \\ \mathbf{l}_2 \end{bmatrix}.$$

Then the surface-wave solution is given by

$$\boldsymbol{u} = \left(c_1 \mathbf{a}_1 + c_2 \left(\mathbf{a}_2 - \sqrt{-1}\,k\,(\mathbf{n}\cdot\mathbf{x})\,\mathbf{a}_1 \right) + c_3 \left(\mathbf{a}_3 - \sqrt{-1}\,k\,(\mathbf{n}\cdot\mathbf{x})\,\mathbf{a}_2 \right. \right.$$

$$\left. \left. - \frac{1}{2}k^2\,(\mathbf{n}\cdot\mathbf{x})^2\,\mathbf{a}_1 \right) \right) e^{-\sqrt{-1}\,k(\mathbf{m}\cdot\mathbf{x}+p_1\mathbf{n}\cdot\mathbf{x}-vt)}, \tag{3.31}$$

and the corresponding traction on the surface $\mathbf{n} \cdot \mathbf{x} = 0$ *produced by the solution above is given by*

$$\boldsymbol{t} = -\sqrt{-1}\,k \sum_{\alpha=1}^{3} c_\alpha \, \mathbf{l}_\alpha \, e^{-\sqrt{-1}\,k(\mathbf{m}\cdot\mathbf{x}-vt)}, \tag{3.32}$$

where c_α $(1 \le \alpha \le 3)$ *are arbitrary complex constants.*

Springer

Although the forms of the surface-wave solutions are different according to the degeneracy of the eigenvalue problem (3.14) (see (3.27), (3.29) and (3.31)), they have the same form on the surface $\mathbf{n} \cdot \mathbf{x} = 0$:

$$\boldsymbol{u} = \sum_{\alpha=1}^{3} c_\alpha \, \mathbf{a}_\alpha \, e^{-\sqrt{-1}\,k(\mathbf{m}\cdot\mathbf{x}-v\,t)}. \tag{3.33}$$

As well, the corresponding tractions on $\mathbf{n} \cdot \mathbf{x} = 0$ all have the same form

$$\boldsymbol{t} = -\sqrt{-1}\,k \sum_{\alpha=1}^{3} c_\alpha \, \mathbf{l}_\alpha \, e^{-\sqrt{-1}\,k(\mathbf{m}\cdot\mathbf{x}-v\,t)} \tag{3.34}$$

(see (3.28), (3.30) and (3.32)). Henceforth, we call the vectors $\mathbf{a}_\alpha \in \mathbb{C}^3$ ($\alpha = 1, 2, 3$) given in each case of the lemma the displacement parts and $\mathbf{l}_\alpha \in \mathbb{C}^3$ ($\alpha = 1, 2, 3$) the traction parts of eigenvectors or generalized eigenvectors $\begin{bmatrix} \mathbf{a}_\alpha \\ \mathbf{l}_\alpha \end{bmatrix}$ ($\alpha = 1, 2, 3$) of $\mathbf{N}$.

Example Isotropic elasticity falls under the semisimple case for any $\mathbf{m}$ and $\mathbf{n}$ when $0 < v < v_L^{\text{Iso}} = \sqrt{\frac{\mu}{\rho}}$. In fact, $\mathbf{Q} + p(\mathbf{R} + \mathbf{R}^T) + p^2 \mathbf{T}$ in (3.10) becomes the matrix which is obtained by adding the term $-\rho v^2$ to the diagonal components of the matrix (1.82), and it can be checked from (3.11) that

$$p_1 = p_2 = \sqrt{-1}\sqrt{\frac{\mu - \rho v^2}{\mu}}, \qquad p_3 = \sqrt{-1}\sqrt{\frac{\lambda + 2\mu - \rho v^2}{\lambda + 2\mu}}. \tag{3.35}$$

Since the strong convexity condition (1.83) implies that $\mu < \lambda + 2\mu$, we see from (2) of Lemma 3.2 that $v_L^{\text{Iso}} = \sqrt{\frac{\mu}{\rho}}$. The corresponding displacement parts $\mathbf{a}_\alpha$ ($\alpha = 1, 2, 3$) of eigenvectors of $\mathbf{N}$ in isotropic elasticity can be taken as

$$\mathbf{a}_1 = \mathbf{m} \times \mathbf{n}, \qquad \mathbf{a}_2 = \mathbf{n} - p_1 \mathbf{m}, \qquad \mathbf{a}_3 = \mathbf{m} + p_3 \mathbf{n} \tag{3.36}$$

(Exercise 3-1). The first two vectors are linearly independent. Therefore, the eigenvalue problem (3.14) for isotropic elasticity is semisimple for $0 < v < v_L^{\text{Iso}}$. Note that when $v = 0$, the eigenvalue problem (3.14) is degenerate and the length of the Jordan chain is two (Subsection 1.8.1).

Now for $0 \le v < v_L$, let $\mathbf{N}(\phi)$ be the 6×6 real matrix defined by

$$\mathbf{N}(\phi) = \begin{bmatrix} -\mathbf{T}(\phi)^{-1}\mathbf{R}(\phi)^T & \mathbf{T}(\phi)^{-1} \\ -\mathbf{Q}(\phi) + \mathbf{R}(\phi)\mathbf{T}(\phi)^{-1}\mathbf{R}(\phi)^T & -\mathbf{R}(\phi)\mathbf{T}(\phi)^{-1} \end{bmatrix}, \tag{3.37}$$

where $\mathbf{Q}(\phi), \mathbf{R}(\phi)$ and $\mathbf{T}(\phi)$ are given by (3.17). Then $\mathbf{N}(0)$ equals $\mathbf{N}$ in (3.15). To consider how the eigenvalues, eigenvectors and generalized eigenvectors of $\mathbf{N}(\phi)$ depend on the angle ϕ, we recall that the major symmetry (3.18) of the dynamic elastic coefficients implies that

$$\mathbf{R}(\phi)^T = \left(\sum_{j,l=1}^{3} C_{ijkl}^{d}\, \tilde{n}_j \tilde{m}_l \right)_{i\downarrow k \rightarrow 1,2,3}.$$

Then the same differentiation formulas as (1.35) hold for $\mathbf{Q}(\phi)$, $\mathbf{R}(\phi)$ and $\mathbf{T}(\phi)$ in (3.17). Hence, when $0 \leq v < v_L$, we obtain for dynamic elasticity the same results on rotational dependence as Theorem 1.5, Theorem 1.9 and Theorem 1.10.

However, in these results, eigenvectors and generalized eigenvectors of $\mathbf{N}(\phi)$ no longer have physical meaning when $\phi \neq 0$,[34] because of the term $\rho v^2 m_j m_l \delta_{ik}$ in (3.6). But, we need these results in order to derive the Barnett-Lothe integral formalism for dynamic elasticity, to which we now proceed.

Definition 3.4 For $0 \leq v < v_L$, we define the 6×6 real matrix $\mathbf{S} = \mathbf{S}(v)$ to be the angular average of the 6×6 matrix $\mathbf{N}(\phi)$ over $[-\pi, \pi]$:

$$\mathbf{S} = \begin{bmatrix} \mathbf{S}_1 & \mathbf{S}_2 \\ \mathbf{S}_3 & \mathbf{S}_1^T \end{bmatrix} = \frac{1}{2\pi} \int_{-\pi}^{\pi} \mathbf{N}(\phi)\, d\phi, \tag{3.38}$$

where $\mathbf{S}_1 = \mathbf{S}_1(v)$, $\mathbf{S}_2 = \mathbf{S}_2(v)$ and $\mathbf{S}_3 = \mathbf{S}_3(v)$ are 3×3 real matrices defined by

$$\mathbf{S}_1 = \frac{1}{2\pi} \int_{-\pi}^{\pi} -\mathbf{T}(\phi)^{-1}\mathbf{R}(\phi)^T\, d\phi, \qquad \mathbf{S}_2 = \frac{1}{2\pi} \int_{-\pi}^{\pi} \mathbf{T}(\phi)^{-1}\, d\phi,$$

$$\mathbf{S}_3 = \frac{1}{2\pi} \int_{-\pi}^{\pi} -\mathbf{Q}(\phi) + \mathbf{R}(\phi)\mathbf{T}(\phi)^{-1}\mathbf{R}(\phi)^T\, d\phi \tag{3.39}$$

and $\mathbf{Q}(\phi)$, $\mathbf{R}(\phi)$ and $\mathbf{T}(\phi)$ are given by (3.17).

From (1) of Lemma 3.2 and from the proof of Lemma 1.14 we immediately arrive at

Lemma 3.5 *For $0 \leq v < v_L$, the matrices $\mathbf{S}_2$ and $\mathbf{S}_3$ are symmetric. Furthermore, $\mathbf{S}_2$ is positive definite.*

Now we take the angular average of Stroh's eigenvalue problem as in the proof of Theorem 1.15 to obtain the fundamental theorem in the Barnett-Lothe integral formalism for dynamic elasticity. Thus we have

Theorem 3.6 *For $0 \leq v < v_L$, let $\begin{bmatrix} \mathbf{a}_\alpha \\ \mathbf{l}_\alpha \end{bmatrix}$ be an eigenvector or generalized eigenvector of $\mathbf{N}(0)$ corresponding to the eigenvalues p_α ($\alpha = 1, 2, 3$) with $\mathrm{Im}\, p_\alpha > 0$. Then for $0 \leq v < v_L$,*

$$\mathbf{S} \begin{bmatrix} \mathbf{a}_\alpha \\ \mathbf{l}_\alpha \end{bmatrix} = \sqrt{-1} \begin{bmatrix} \mathbf{a}_\alpha \\ \mathbf{l}_\alpha \end{bmatrix}. \tag{3.40}$$

---

[34]In Chapter 1, eigenvectors and generalized eigenvectors of $\mathbf{N}(\phi)$ have the physical significance that they are composed of the displacement parts $\mathbf{a}(\phi)$ and the traction parts $\mathbf{l}(\phi)$ on the elastic half-space rotated around $\mathbf{m} \times \mathbf{n}$ by the angle ϕ.

Then, as in the proof of Theorem 1.16, from Lemma 3.5 and Theorem 3.6 we obtain

Corollary 3.7 *For* $0 \leq v < v_L$, *let* $\begin{bmatrix} \mathbf{a}_\alpha \\ \mathbf{l}_\alpha \end{bmatrix}$ $(\alpha = 1, 2, 3)$ *be linearly independent eigenvector(s) or generalized eigenvector(s) of* $\mathbf{N}(0)$ *corresponding to the eigenvalues* p_α $(\alpha = 1, 2, 3)$ *with* $\operatorname{Im} p_\alpha > 0$. *Then their displacement parts* $\mathbf{a}_\alpha$ $(\alpha = 1, 2, 3)$ *are linearly independent.*

Thus, we can define the surface impedance matrix for dynamic elasticity.

Definition 3.8 For $0 \leq v < v_L$, the surface impedance matrix $\mathbf{Z} = \mathbf{Z}(v)$ is the 3×3 matrix given by

$$\mathbf{Z} = -\sqrt{-1}\,[\mathbf{l}_1, \mathbf{l}_2, \mathbf{l}_3][\mathbf{a}_1, \mathbf{a}_2, \mathbf{a}_3]^{-1}, \tag{3.41}$$

where $[\mathbf{l}_1, \mathbf{l}_2, \mathbf{l}_3]$ and $[\mathbf{a}_1, \mathbf{a}_2, \mathbf{a}_3]$ denote 3×3 matrices which consist of the column vectors $\mathbf{l}_\alpha$ and $\mathbf{a}_\alpha$ respectively.

Therefore, $\mathbf{Z}$ expresses a linear relationship between (i) the displacements at the surface $\mathbf{n} \cdot \mathbf{x} = 0$ on which surface waves of the form (3.27), (3.29) or (3.31) propagate in the direction of $\mathbf{m}$ with the phase velocity v and (ii) the tractions needed to sustain them at that surface.

As in the derivation of Theorem 1.18, from (3.38), (3.40) and (3.41) we obtain the integral representation of $\mathbf{Z}$.

Theorem 3.9 *For* $0 \leq v < v_L$,

$$\mathbf{Z} = \mathbf{S}_2^{-1} + \sqrt{-1}\,\mathbf{S}_2^{-1}\mathbf{S}_1, \tag{3.42}$$

where the matrices $\mathbf{S}_1$ *and* $\mathbf{S}_2$ *are given by* (3.39).

As in the proof of Lemma 1.22, we get from Theorem 3.6 that

$$\mathbf{S}^2 = -\mathbf{I} \qquad (0 \leq v < v_L), \tag{3.43}$$

where $\mathbf{I}$ denotes the 6×6 identity matrix. Then the blockwise expression of (3.43) as obtained from (3.38) gives

$$\mathbf{S}_1^2 + \mathbf{S}_2\mathbf{S}_3 = -\mathbf{I}, \qquad \mathbf{S}_1\mathbf{S}_2 + \mathbf{S}_2\mathbf{S}_1^T = \mathbf{O} \qquad (0 \leq v < v_L). \tag{3.44}$$

Since $\mathbf{S}_2$ is symmetric and invertible for $0 \leq v < v_L$, it follows from the second equality in (3.44) that

$$\mathbf{S}_2^{-1}\mathbf{S}_1 = -\left(\mathbf{S}_2^{-1}\mathbf{S}_1\right)^T \qquad (0 \leq v < v_L), \tag{3.45}$$

which implies that $\mathbf{S}_2^{-1}\mathbf{S}_1$ is anti-symmetric. Hence we obtain

Corollary 3.10 *The surface impedance matrix* $\mathbf{Z}$ *is Hermitian for* $0 \leq v < v_L$.

It is proved that $\mathbf{Z}$ is positive definite at $v = 0$ (see the comment after Theorem 1.21). As we shall see later, the eigenvalues of $\mathbf{Z}(v)$ decrease monotonically with v in the interval $0 \leq v < v_L$ from their positive values at $v = 0$ (Lemma 3.33), and one of them becomes zero when v equals the Rayleigh-wave velocity v_R (Theorem 3.13).

3.3 Rayleigh Waves in Elastic Half-space

Rayleigh waves are elastic surface waves which propagate along the traction-free surface with the phase velocity in the subsonic range, and whose amplitude decays exponentially with depth below that surface. Let $\mathbf{m}$ and $\mathbf{n}$ be orthogonal unit vectors in $\mathbb{R}^3$. Following the setting of Section 3.1, we consider Rayleigh waves which propagate along the surface $\mathbf{n} \cdot \mathbf{x} = 0$ in the direction of $\mathbf{m}$ with the phase velocity v_R satisfying $0 < v_R < v_L$, and whose amplitude decays exponentially as $\mathbf{n} \cdot \mathbf{x} \longrightarrow -\infty$, and which produce no tractions on $\mathbf{n} \cdot \mathbf{x} = 0$ (see (3.27), (3.29) and (3.31)). Here $v_L = v_L(\mathbf{m}, \mathbf{n})$ is the limiting velocity in Definition 3.1.

Elastic surface waves which propagate along the traction-free surface with the phase velocity greater than the limiting velocity are called supersonic surface waves. But we do not consider them in this chapter.[35]

We take $\mathbb{C}^3$-vectors $\mathbf{a}_\alpha$ and $\mathbf{l}_\alpha$ ($\alpha = 1, 2, 3$) so that $\begin{bmatrix} \mathbf{a}_\alpha \\ \mathbf{l}_\alpha \end{bmatrix}$ ($\alpha = 1, 2, 3$) are linearly independent eigenvector(s) or generalized eigenvector(s) of $\mathbf{N} = \mathbf{N}(0)$ at $v = v_R$ associated with the eigenvalues p_α ($\alpha = 1, 2, 3$, $\mathrm{Im}\, p_\alpha > 0$). Then the existence of Rayleigh waves implies that the corresponding traction on $\mathbf{n} \cdot \mathbf{x} = 0$ given by (3.34) vanishes for $v = v_R$. In other words, there exists a set of complex numbers $(c_1, c_2, c_3) \neq (0, 0, 0)$ such that

$$\sum_{\alpha=1}^{3} c_\alpha \mathbf{l}_\alpha = \mathbf{0} \qquad \text{at } v = v_R, \tag{3.46}$$

which is equivalent to

$$\det[\mathbf{l}_1, \mathbf{l}_2, \mathbf{l}_3] = 0 \qquad \text{at } v = v_R. \tag{3.47}$$

If the traction parts $\mathbf{l}_\alpha$ ($\alpha = 1, 2, 3$) are computed explicitly, it will be straightforward to derive the secular equation for v_R from (3.47). For isotropic elasticity, by (3.47), the equation for v_R becomes bicubic (Exercise 3-2).

The next Lemma provides an alternative method for finding the secular equation for v_R, where the integral formalism developed in the last section will be used. Then it is neither necessary to see whether the eigenvalue problem (3.14) is simple, semisimple or degenerate, nor necessary to seek explicit forms of the traction parts $\mathbf{l}_\alpha$ ($\alpha = 1, 2, 3$).

Lemma 3.11 *For given orthogonal unit vectors* $\mathbf{m}$ *and* $\mathbf{n}$ *in* $\mathbb{R}^3$, *let* $\mathbf{S}_3 = \mathbf{S}_3(v)$ *be the* 3×3 *real matrix defined by* (3.39) *for* $0 \leq v < v_L$. *Then a necessary and sufficient condition for the existence of the Rayleigh waves in the half-space* $\mathbf{n} \cdot \mathbf{x} \leq 0$ *which propagate along the surface* $\mathbf{n} \cdot \mathbf{x} = 0$ *in the direction of* $\mathbf{m}$ *with the phase velocity* v_R *in the subsonic range* $0 < v < v_L$ *is*

$$\det \mathbf{S}_3 = 0 \qquad \text{at } v = v_R \qquad (0 < v_R < v_L). \tag{3.48}$$

---

[35] For references on supersonic surface waves, see the first paragraph of Section 3.7.

Proof Suppose that Rayleigh waves exist. The first three rows of the system (3.40) are written, by using notations in (3.38), as

$$S_1 a_\alpha + S_2 l_\alpha = \sqrt{-1}\, a_\alpha, \qquad \alpha = 1, 2, 3 \tag{3.49}$$

and the last three rows of the system (3.40) are written as

$$S_3 a_\alpha + S_1^T l_\alpha = \sqrt{-1}\, l_\alpha, \qquad \alpha = 1, 2, 3. \tag{3.50}$$

Then from (3.46) and (3.50) it follows that for c_α in (3.46),

$$S_3 \sum_{\alpha=1}^{3} c_\alpha a_\alpha = 0 \qquad \text{at } v = v_R. \tag{3.51}$$

Corollary 3.7 implies that

$$\sum_{\alpha=1}^{3} c_\alpha a_\alpha \neq 0 \qquad \text{at } v = v_R.$$

Hence the null space of S_3 is not trivial at $v = v_R$, which gives (3.48).

Conversely, suppose that there is a speed v_R in the subsonic range $0 < v < v_L$ for which (3.48) holds. Let

$$A = [a_1, a_2, a_3], \qquad L = [l_1, l_2, l_3]$$

denote the 3×3 matrices which consist of the displacement parts a_α and the traction parts l_α at $v = v_R$ that we have taken at the beginning of this section. Then (3.49) and (3.50) can be rewritten as

$$S_1 A + S_2 L = \sqrt{-1}\, A$$

and

$$S_3 A + S_1^T L = \sqrt{-1}\, L,$$

respectively. Then

$$S_2 L = \left(\sqrt{-1}\, I - S_1\right) A, \quad S_3 A = \left(\sqrt{-1}\, I - S_1^T\right) L.$$

Taking the determinants of both sides of these equations, we have

$$\det S_2 \det L = \det(\sqrt{-1}\, I - S_1) \det A \tag{3.52}$$

and

$$\det S_3 \det A = \det(\sqrt{-1}\, I - S_1^T) \det L. \tag{3.53}$$

Hence condition (3.48) implies through (3.53) that

$$\det \left(\sqrt{-1}\, I - S_1^T\right) = \det \left(\sqrt{-1}\, I - S_1\right) = 0 \quad \text{or} \quad \det L = 0$$

at $v = v_R$. In the case where $\det(\sqrt{-1}\, I - S_1) = 0$, since Lemma 3.5 guarantees that $\det S_2 \neq 0$ for $0 \leq v < v_L$, from (3.52) we get $\det L = 0$ at $v = v_R$. $\qquad \square$

Furthermore, we shall see that condition (3.48) can be replaced by a more stringent one.

 Springer

Theorem 3.12 *Let* **m** *and* **n** *be orthogonal unit vectors in* $\mathbb{R}^3$. *A necessary and sufficient condition for the existence of the Rayleigh waves in the half-space* $\mathbf{n} \cdot \mathbf{x} \leq 0$ *which propagate along the surface* $\mathbf{n} \cdot \mathbf{x} = 0$ *in the direction of* **m** *with the phase velocity* v_R *in the subsonic range* $0 < v < v_L$ *is*

$$\text{rank } \mathbf{S}_3 = 1 \quad at \ v = v_R \quad (0 < v_R < v_L). \tag{3.54}$$

Proof We only have to prove necessity. Let $\mathbf{a}^+$ and $\mathbf{a}^-$ be the real part and the imaginary part of the vector $\sum_{\alpha=1}^{3} c_\alpha \mathbf{a}_\alpha$ at $v = v_R$, respectively:

$$\sum_{\alpha=1}^{3} c_\alpha \mathbf{a}_\alpha = \mathbf{a}^+ + \sqrt{-1}\,\mathbf{a}^-, \qquad \mathbf{a}^+, \mathbf{a}^- \in \mathbb{R}^3,$$

where c_α ($\alpha = 1, 2, 3$) are the coefficients that satisfy (3.46).

We will show that $\mathbf{a}^+$ and $\mathbf{a}^-$ are linearly independent. Then, by (3.51), the null space of $\mathbf{S}_3$ is two-dimensional, which gives (3.54).

From (3.46) and (3.49) it follows that

$$\mathbf{S}_1 \left(\mathbf{a}^+ + \sqrt{-1}\,\mathbf{a}^- \right) = \sqrt{-1} \left(\mathbf{a}^+ + \sqrt{-1}\,\mathbf{a}^- \right) \qquad at \ v = v_R.$$

Since the matrix $\mathbf{S}_1$ is real, we get

$$\mathbf{S}_1 \mathbf{a}^+ = -\mathbf{a}^- \quad and \quad \mathbf{S}_1 \mathbf{a}^- = \mathbf{a}^+. \tag{3.55}$$

We have $\mathbf{a}^+ \neq \mathbf{0}$ and $\mathbf{a}^- \neq \mathbf{0}$; otherwise from (3.55) it follows that $\sum_{\alpha=1}^{3} c_\alpha \mathbf{a}_\alpha = \mathbf{0}$, which contradicts Corollary 3.7. Suppose that there exists a real number k such that $\mathbf{a}^+ = k\,\mathbf{a}^-$ holds. Then by (3.55),

$$-\mathbf{a}^- = \mathbf{S}_1 \mathbf{a}^+ = \mathbf{S}_1(k\mathbf{a}^-) = k\,\mathbf{S}_1 \mathbf{a}^- = k\,\mathbf{a}^+ = k^2 \mathbf{a}^-,$$

which implies $k^2 = -1$. This is a contradiction.

Suppose that rank $\mathbf{S}_3 = 0$ at $v = v_R$. Then $\mathbf{S}_3$ vanishes there and the first equality in (3.44) gives $\mathbf{S}_1^2 = -\mathbf{I}$ at $v = v_R$. Since $\mathbf{S}_2$ is invertible for $0 \leq v < v_L$, taking the determinants of both sides of (3.45), we have $\det \mathbf{S}_1 = 0$ for $0 \leq v < v_L$. This is a contradiction. □

We give another characterization of the Rayleigh-wave velocity v_R in terms of the surface impedance matrix $\mathbf{Z} = \mathbf{Z}(v)$ in Definition 3.8.

Theorem 3.13 *A necessary and sufficient condition for the existence of Rayleigh waves in Theorem 3.12 is*

$$\det \mathbf{Z} = 0 \quad at \ v = v_R \quad (0 < v_R < v_L). \tag{3.56}$$

Proof Obvious from Corollary 3.7, (3.41) and (3.47). □

In the forthcoming Sections 3.4 and 3.5 we shall use Theorem 3.12 to find the secular equations for v_R in isotropic elasticity and in weakly anisotropic elastic media. At this point, we prove that v_R is unique if it exists. That is, we have

🍃 Springer

Theorem 3.14 *For given orthogonal unit vectors* **m** *and* **n**, *the Rayleigh wave is unique if it exists.*

This theorem is proved by further consideration on the matrix $\mathbf{S}_3$.

Lemma 3.15 *For* $0 \le v < v_L$, *let* $\mathbf{S}_3 = \mathbf{S}_3(v)$ *be the* 3×3 *real symmetric matrix in* (3.39).

(1)   *The symmetric matrix* $\frac{d}{dv}\mathbf{S}_3$ *is positive definite.*
(2)   *The eigenvalues of* $\mathbf{S}_3$ *are monotonic increasing functions of* v.

Proof To prove (1), we first differentiate the integrand of $\mathbf{S}_3$, namely

$$-\mathbf{Q}(\phi) + \mathbf{R}(\phi)\mathbf{T}(\phi)^{-1}\mathbf{R}(\phi)^T, \tag{3.57}$$

with respect to v. From (3.6), (3.8), (3.16) and (3.17) it follows that

$$\mathbf{Q}(\phi) = \left(\sum_{j,l=1}^{3} C_{ijkl}\,\tilde{m}_j\tilde{m}_l\right)_{i\downarrow k\to 1,2,3} - V\left(\sum_{j,l=1}^{3} m_jm_l\,\delta_{ik}\,\tilde{m}_j\tilde{m}_l\right)_{i\downarrow k\to 1,2,3}$$

$$= \left(\sum_{j,l=1}^{3} C_{ijkl}\,\tilde{m}_j\tilde{m}_l\right)_{i\downarrow k\to 1,2,3} - V\cos^2\phi\,\mathbf{I},$$

$$\mathbf{R}(\phi) = \left(\sum_{j,l=1}^{3} C_{ijkl}\,\tilde{m}_j\tilde{n}_l\right)_{i\downarrow k\to 1,2,3} - V\left(\sum_{j,l=1}^{3} m_jm_l\,\delta_{ik}\,\tilde{m}_j\tilde{n}_l\right)_{i\downarrow k\to 1,2,3}$$

$$= \left(\sum_{j,l=1}^{3} C_{ijkl}\,\tilde{m}_j\tilde{n}_l\right)_{i\downarrow k\to 1,2,3} + V\cos\phi\sin\phi\,\mathbf{I},$$

$$\mathbf{T}(\phi) = \left(\sum_{j,l=1}^{3} C_{ijkl}\,\tilde{n}_j\tilde{n}_l\right)_{i\downarrow k\to 1,2,3} - V\left(\sum_{j,l=1}^{3} m_jm_l\,\delta_{ik}\,\tilde{n}_j\tilde{n}_l\right)_{i\downarrow k\to 1,2,3}$$

$$= \left(\sum_{j,l=1}^{3} C_{ijkl}\,\tilde{n}_j\tilde{n}_l\right)_{i\downarrow k\to 1,2,3} - V\sin^2\phi\,\mathbf{I}, \tag{3.58}$$

where $V = \rho v^2$. Then we have

$$\frac{d}{dv}\mathbf{Q}(\phi) = -2\rho\,v\cos^2\phi\,\mathbf{I}, \qquad \frac{d}{dv}\mathbf{R}(\phi) = 2\rho\,v\cos\phi\sin\phi\,\mathbf{I} = \frac{d}{dv}\mathbf{R}(\phi)^T.$$

Also, since

$$\frac{d}{dv}\mathbf{T}(\phi) = -2\rho\,v\sin^2\phi\,\mathbf{I}$$

and

$$\frac{d}{dv}\left(\mathbf{T}(\phi)\mathbf{T}(\phi)^{-1}\right) = \frac{d}{dv}\mathbf{T}(\phi)\cdot\mathbf{T}(\phi)^{-1} + \mathbf{T}(\phi)\frac{d}{dv}\mathbf{T}(\phi)^{-1} = \mathbf{0},$$

we get

$$\frac{d}{dv}\mathbf{T}(\phi)^{-1} = -\mathbf{T}(\phi)^{-1}\frac{d}{dv}\mathbf{T}(\phi) \cdot \mathbf{T}(\phi)^{-1} = 2\rho\, v \sin^2\phi\, \mathbf{T}(\phi)^{-2}.$$

Therefore, the derivative of the integrand (3.57) in v becomes

$$\frac{d}{dv}\left[-\mathbf{Q}(\phi) + \mathbf{R}(\phi)\mathbf{T}(\phi)^{-1}\mathbf{R}(\phi)^T\right]$$

$$= 2\rho\, v\left[\cos^2\phi\, \mathbf{I} + \cos\phi\sin\phi\left(\mathbf{T}(\phi)^{-1}\mathbf{R}(\phi)^T + \mathbf{R}(\phi)\mathbf{T}(\phi)^{-1}\right)\right.$$
$$\left. + \sin^2\phi\, \mathbf{R}(\phi)\mathbf{T}(\phi)^{-2}\mathbf{R}(\phi)^T\right].$$

We note that the right hand side of the preceding equation can be written as

$$2\rho\, v\left[\cos\phi\, \mathbf{I} + \sin\phi\, \mathbf{T}(\phi)^{-1}\mathbf{R}(\phi)^T\right]^T\left[\cos\phi\, \mathbf{I} + \sin\phi\, \mathbf{T}(\phi)^{-1}\mathbf{R}(\phi)^T\right].$$

Then for any non-zero $\mathbf{v} \in \mathbb{R}^3$ it follows that

$$\mathbf{v}\cdot\left(\frac{d\mathbf{S}_3}{dv}\mathbf{v}\right) = \frac{1}{2\pi}\int_{-\pi}^{\pi}\mathbf{v}\cdot\frac{d}{dv}\left[-\mathbf{Q}(\phi) + \mathbf{R}(\phi)\mathbf{T}(\phi)^{-1}\mathbf{R}(\phi)^T\right]\mathbf{v}\, d\phi$$

$$= \frac{\rho\, v}{\pi}\int_{-\pi}^{\pi}\left[(\cos\phi\, \mathbf{I} + \sin\phi\, \mathbf{T}(\phi)^{-1}\mathbf{R}(\phi)^T)\,\mathbf{v}\right]$$
$$\cdot\left[(\cos\phi\, \mathbf{I} + \sin\phi\, \mathbf{T}(\phi)^{-1}\mathbf{R}(\phi)^T)\,\mathbf{v}\right]d\phi,$$

which is positive, because the last integrand is non-negative for all ϕ and is strictly positive for $\phi = 0$. This proves (1).

Let $\mathbf{v}_i = \mathbf{v}_i(v) \in \mathbb{R}^3$ $(i = 1, 2, 3)$ be eigenvectors of the real symmetric matrix $\mathbf{S}_3$ associated with the eigenvalues $\mu_i = \mu_i(v) \in \mathbb{R}$ $(i = 1, 2, 3)$. Differentiating the eigenrelations

$$\mathbf{S}_3\mathbf{v}_i = \mu_i\mathbf{v}_i, \qquad i = 1, 2, 3 \qquad\qquad (3.59)$$

with respect to v, we have

$$\frac{d\mathbf{S}_3}{dv}\mathbf{v}_i + \mathbf{S}_3\frac{d\mathbf{v}_i}{dv} = \frac{d\mu_i}{dv}\mathbf{v}_i + \mu_i\frac{d\mathbf{v}_i}{dv}, \qquad i = 1, 2, 3.$$

Taking the inner products with $\mathbf{v}_i$, we get

$$\mathbf{v}_i\cdot\left(\frac{d\mathbf{S}_3}{dv}\mathbf{v}_i\right) + \mathbf{v}_i\cdot\left(\mathbf{S}_3\frac{d\mathbf{v}_i}{dv}\right) = \frac{d\mu_i}{dv}\mathbf{v}_i\cdot\mathbf{v}_i + \mu_i\,\mathbf{v}_i\cdot\frac{d\mathbf{v}_i}{dv}, \qquad i = 1, 2, 3.$$

Since $\mathbf{S}_3$ is symmetric, by (3.59) we have

$$\mathbf{v}_i\cdot\left(\mathbf{S}_3\frac{d\mathbf{v}_i}{dv}\right) = (\mathbf{S}_3\mathbf{v}_i)\cdot\frac{d\mathbf{v}_i}{dv} = \mu_i\mathbf{v}_i\cdot\frac{d\mathbf{v}_i}{dv}, \qquad i = 1, 2, 3.$$

Then it follows that

$$\mathbf{v}_i\cdot\left(\frac{d\mathbf{S}_3}{dv}\mathbf{v}_i\right) = \frac{d\mu_i}{dv}\mathbf{v}_i\cdot\mathbf{v}_i, \qquad i = 1, 2, 3.$$

By (1) of the lemma, the left hand side is positive. Also, we have $\mathbf{v}_i \cdot \mathbf{v}_i > 0$ ($i = 1, 2, 3$). Thus,

$$\frac{d\mu_i}{dv} > 0, \qquad i = 1, 2, 3.$$

This proves (2).   □

Proof of Theorem 3.14 By Theorem 3.12, the matrix $\mathbf{S}_3(v_R)$ has a two-dimensional eigenspace corresponding to the multiple zero eigenvalue. This implies that two of the eigenvalues of $\mathbf{S}_3(v_R)$ must vanish together at $v = v_R$. If there were two Rayleigh waves, i.e., there were two v_R's which satisfy (3.54), then one of the three eigenvalues of $\mathbf{S}_3(v)$ would have to vanish twice for $0 \le v < v_L$. This contradicts the monotonicity of the eigenvalues in the last lemma.   □

Finally, we show that the integral formalism can also be applied to get the formula for the polarization vector of the Rayleigh waves at the surface. Let $\begin{bmatrix} \mathbf{a}_\alpha \\ \mathbf{l}_\alpha \end{bmatrix}$ ($\alpha = 1, 2, 3$) be linearly independent eigenvector(s) or generalized eigenvector(s) of $\mathbf{N} = \mathbf{N}(0)$ at $v = v_R$ associated with the eigenvalues p_α ($\alpha = 1, 2, 3$, Im $p_\alpha > 0$). By (3.33), the displacement field of the Rayleigh waves at the surface $\mathbf{n} \cdot \mathbf{x} = 0$ is written as

$$\boldsymbol{u} = \sum_{\alpha=1}^{3} c_\alpha \, \mathbf{a}_\alpha \, e^{-\sqrt{-1}\, k(\mathbf{m}\cdot\mathbf{x} - v_R t)}, \tag{3.60}$$

where c_α ($\alpha = 1, 2, 3$) are the coefficients which satisfy (3.46). We call the vector

$$\mathbf{a}_{\mathrm{pol}} = \sum_{\alpha=1}^{3} c_\alpha \, \mathbf{a}_\alpha \in \mathbb{C}^3$$

the polarization vector of the Rayleigh waves at the surface. This means that the real part $\mathbf{a}^+$ and the imaginary part $\mathbf{a}^-$ of $\mathbf{a}_{\mathrm{pol}}$ define the plane to which the paths of surface particles are confined, so that their displacements are expressed by taking the real part of $\mathbf{a}_{\mathrm{pol}} e^{-\sqrt{-1}\, k(\mathbf{m}\cdot\mathbf{x} - v_R t)}$ or

$$\mathbf{a}^+ \cos k(\mathbf{m} \cdot \mathbf{x} - v_R t) + \mathbf{a}^- \sin k(\mathbf{m} \cdot \mathbf{x} - v_R t). \tag{3.61}$$

Note that the imaginary part of $\mathbf{a}_{\mathrm{pol}} e^{-\sqrt{-1}\, k(\mathbf{m}\cdot\mathbf{x} - v_R t)}$ differs from the real part (3.61) only by a phase shift of 90 degrees.

Now we give an integral expression for the polarization vector $\mathbf{a}_{\mathrm{pol}}$.

Theorem 3.16 *Let $\mathbf{S}_1(v)$ and $\mathbf{S}_3(v)$ be the 3×3 real matrices in (3.39), and let $\mathbf{e}_1$ and $\mathbf{e}_2$ be orthogonal unit vectors in $\mathbb{R}^3$ such that*

$$\mathbf{S}_3(v_R)\mathbf{e}_1 \neq \mathbf{0}, \qquad (\mathbf{S}_3(v_R)\mathbf{e}_1) \times \mathbf{e}_2 \neq \mathbf{0}, \tag{3.62}$$

where the symbol $\times$ denotes the cross product of vectors. Then the Rayleigh waves described in Theorem 3.12 have at the surface $\mathbf{n} \cdot \mathbf{x} = 0$ the polarization vector

$$\mathbf{a}_{\mathrm{pol}} = (\mathbf{S}_3(v_R)\mathbf{e}_1) \times \mathbf{e}_2 - \sqrt{-1}\, \mathbf{S}_1(v_R) \left[(\mathbf{S}_3(v_R)\mathbf{e}_1) \times \mathbf{e}_2 \right]. \tag{3.63}$$

Proof Formula (3.54) implies that there is a non-zero vector $\mathbf{s} \in \mathbb{R}^3$ such that

$$\mathbf{S}_3(v_R) = \pm \mathbf{s} \otimes \mathbf{s}, \tag{3.64}$$

where the symbol $\otimes$ denotes the tensor product of two vectors.[36] Then from (3.51) it follows that

$$\mathbf{S}_3(v_R)\mathbf{a}_{\mathrm{pol}} = \pm(\mathbf{s} \otimes \mathbf{s})\mathbf{a}_{\mathrm{pol}} = \pm(\mathbf{a}_{\mathrm{pol}}, \mathbf{s})_{\mathbb{C}^3}\,\mathbf{s} = \mathbf{0}. \tag{3.66}$$

Since $\mathbf{s} \neq \mathbf{0}$, we have

$$\left(\mathbf{a}_{\mathrm{pol}}, \mathbf{s}\right)_{\mathbb{C}^3} = \mathbf{a}^+ \cdot \mathbf{s} + \sqrt{-1}\,\mathbf{a}^- \cdot \mathbf{s} = 0,$$

and hence

$$\mathbf{a}^+ \cdot \mathbf{s} = \mathbf{a}^- \cdot \mathbf{s} = 0.$$

This implies that $\mathbf{s}$ is normal to the plane of polarization.

Let $\mathbf{e}_1$ and $\mathbf{e}_2$ be orthogonal unit vectors in $\mathbb{R}^3$ which satisfiy (3.62). Then $\mathbf{S}_3(v_R)\mathbf{e}_1$, which is equal to $(\mathbf{e}_1 \cdot \mathbf{s})\mathbf{s}$ or $-(\mathbf{e}_1 \cdot \mathbf{s})\mathbf{s}$, is a non-zero scalar multiple of $\mathbf{s}$, and hence, is normal to the plane of polarization. Therefore,

$$(\mathbf{S}_3(v_R)\mathbf{e}_1) \times \mathbf{e}_2$$

is a non-zero real vector lying in the plane of polarization. By taking this vector to be a real part $\mathbf{a}^+$, from (3.55) the imaginary part $\mathbf{a}^-$ is given by

$$-\mathbf{S}_1(v_R)\left[(\mathbf{S}_3(v_R)\mathbf{e}_1) \times \mathbf{e}_2\right].$$

$\square$

3.4 Rayleigh Waves in Isotropic Elasticity

In this section, on the basis of Theorem 3.12 and Theorem 3.16, we derive the secular equation for the phase velocity v_R^{Iso} and compute the polarization vector $\mathbf{a}_{\mathrm{pol}}^{\mathrm{Iso}}$ of Rayleigh waves which propagate along the surface of an isotropic elastic half-space.

We first compute the integrand of $\mathbf{S}_3(v)$, namely

$$-\mathbf{Q}(\phi) + \mathbf{R}(\phi)\mathbf{T}(\phi)^{-1}\mathbf{R}(\phi)^T,$$

for which we use formulas (3.58). Since the components of the elasticity tensor of an isotropic material is given by

$$C_{ijkl} = \lambda\,\delta_{ij}\delta_{kl} + \mu(\delta_{ik}\delta_{jl} + \delta_{il}\delta_{kj}), \tag{3.67}$$

---

[36]The tensor product $\mathbf{a} \otimes \mathbf{b}$ of two vectors $\mathbf{a} = (a_1, a_2, a_3)$ and $\mathbf{b} = (b_1, b_2, b_3)$ in $\mathbb{R}^3$ (resp. $\mathbb{C}^3$) can be expressed as a 3×3 matrix

$$\mathbf{a} \otimes \mathbf{b} = (a_i\,b_j)_{i\downarrow j\rightarrow 1,2,3} \qquad \left(\text{resp. } \mathbf{a} \otimes \mathbf{b} = \left(a_i\,\overline{b}_j\right)_{i\downarrow j\rightarrow 1,2,3}\right).$$

Then it follows that

$$(\mathbf{a} \otimes \mathbf{b})\mathbf{c} = (\mathbf{c} \cdot \mathbf{b})\mathbf{a} \qquad (\text{resp. } (\mathbf{a} \otimes \mathbf{b})\mathbf{c} = (\mathbf{c}, \mathbf{b})_{\mathbb{C}^3}\,\mathbf{a}) \tag{3.65}$$

for all $\mathbf{c} \in \mathbb{R}^3$ (resp. $\mathbf{c} \in \mathbb{C}^3$). This will be used in (3.66). Formula (3.64) can be proved by noting from (3.54) that the 3×3 matrix $\mathbf{S}_3(v_R)$ has only one linearly independent column vector.

 Springer

from the identities

$$\widetilde{\mathbf{m}} \cdot \widetilde{\mathbf{m}} = \sum_{j=1}^{3} \widetilde{m}_j^2 = 1, \qquad \widetilde{\mathbf{m}} \cdot \widetilde{\mathbf{n}} = \sum_{j=1}^{3} \widetilde{m}_j \widetilde{n}_j = 0, \qquad \widetilde{\mathbf{n}} \cdot \widetilde{\mathbf{n}} = \sum_{j=1}^{3} \widetilde{n}_j^2 = 1,$$

we have

$$\left(\sum_{j,l=1}^{3} C_{ijkl} \, \widetilde{m}_j \widetilde{m}_l \right)_{i \downarrow k \to 1,2,3} = \left(\lambda \, \widetilde{m}_i \widetilde{m}_k + \mu \big(\delta_{ik} + \widetilde{m}_i \widetilde{m}_k \big) \right)_{i \downarrow k \to 1,2,3}$$

$$= (\lambda + \mu) \, \widetilde{\mathbf{m}} \otimes \widetilde{\mathbf{m}} + \mu \, \mathbf{I},$$

$$\left(\sum_{j,l=1}^{3} C_{ijkl} \, \widetilde{m}_j \widetilde{n}_l \right)_{i \downarrow k \to 1,2,3} = \left(\lambda \, \widetilde{m}_i \widetilde{n}_k + \mu \widetilde{n}_i \widetilde{m}_k \right)_{i \downarrow k \to 1,2,3}$$

$$= \lambda \, \widetilde{\mathbf{m}} \otimes \widetilde{\mathbf{n}} + \mu \, \widetilde{\mathbf{n}} \otimes \widetilde{\mathbf{m}},$$

$$\left(\sum_{j,l=1}^{3} C_{ijkl} \, \widetilde{n}_j \widetilde{n}_l \right)_{i \downarrow k \to 1,2,3} = \left(\lambda \, \widetilde{n}_i \widetilde{n}_k + \mu \big(\delta_{ik} + \widetilde{n}_i \widetilde{n}_k \big) \right)_{i \downarrow k \to 1,2,3}$$

$$= (\lambda + \mu) \, \widetilde{\mathbf{n}} \otimes \widetilde{\mathbf{n}} + \mu \, \mathbf{I}, \tag{3.68}$$

where $\mathbf{I}$ is the 3×3 identity matrix. Hence,

$$\mathbf{Q}(\phi) = (\lambda + \mu) \, \widetilde{\mathbf{m}} \otimes \widetilde{\mathbf{m}} + (\mu - V \cos^2 \phi) \, \mathbf{I}, \tag{3.69}$$

$$\mathbf{R}(\phi) = \lambda \, \widetilde{\mathbf{m}} \otimes \widetilde{\mathbf{n}} + \mu \, \widetilde{\mathbf{n}} \otimes \widetilde{\mathbf{m}} + V \cos \phi \sin \phi \, \mathbf{I}, \tag{3.70}$$

$$\mathbf{T}(\phi) = (\lambda + \mu) \, \widetilde{\mathbf{n}} \otimes \widetilde{\mathbf{n}} + (\mu - V \sin^2 \phi) \, \mathbf{I}, \tag{3.71}$$

where $V = \rho \, v^2$. Let

$$\boldsymbol{\ell} = \widetilde{\mathbf{m}} \times \widetilde{\mathbf{n}} = \mathbf{m} \times \mathbf{n}.$$

Then the triad $\{\widetilde{\mathbf{m}}, \widetilde{\mathbf{n}}, \boldsymbol{\ell}\}$ forms an orthonormal basis in $\mathbb{R}^3$, and we have

$$\mathbf{I} = \widetilde{\mathbf{m}} \otimes \widetilde{\mathbf{m}} + \widetilde{\mathbf{n}} \otimes \widetilde{\mathbf{n}} + \boldsymbol{\ell} \otimes \boldsymbol{\ell}. \tag{3.72}$$

To compute $\mathbf{T}(\phi)^{-1}$, we use (3.72) and rewrite (3.71) as

$$\mathbf{T}(\phi) = (\mu - V \sin^2 \phi) \, (\widetilde{\mathbf{m}} \otimes \widetilde{\mathbf{m}} + \boldsymbol{\ell} \otimes \boldsymbol{\ell}) + (\lambda + 2\mu - V \sin^2 \phi) \, \widetilde{\mathbf{n}} \otimes \widetilde{\mathbf{n}}.^{37}$$

Then by (2.72) we get

$$\mathbf{T}(\phi)^{-1} = \frac{1}{\mu - V \sin^2 \phi} \, (\widetilde{\mathbf{m}} \otimes \widetilde{\mathbf{m}} + \boldsymbol{\ell} \otimes \boldsymbol{\ell}) + \frac{1}{\lambda + 2\mu - V \sin^2 \phi} \, \widetilde{\mathbf{n}} \otimes \widetilde{\mathbf{n}}. \tag{3.73}$$

---

[37] It follows immediately that

$$\det \mathbf{T}(\phi) = (\mu - V \sin^2 \phi)^2 (\lambda + 2\mu - V \sin^2 \phi).$$

From the strong convexity condition (1.83) we get $\mu < \lambda + 2\mu$. Therefore from (3.20) we obtain $v_L^{\text{Iso}} = \sqrt{\frac{\mu}{\rho}}$.

Note that

$$0 \leq V (= \rho\, v^2) \ < \mu < \lambda + 2\mu \tag{3.74}$$

for $0 \leq v < v_L^{\text{Iso}}$. Now from (3.73) and

$$\mathbf{R}(\phi)^T = \lambda\, \tilde{\mathbf{n}} \otimes \tilde{\mathbf{m}} + \mu\, \tilde{\mathbf{m}} \otimes \tilde{\mathbf{n}} + V \cos\phi \sin\phi\, \mathbf{I},$$

it follows that[38]

$$\mathbf{T}(\phi)^{-1}\mathbf{R}(\phi)^T = \frac{\mu}{\mu - V \sin^2\phi}\, \tilde{\mathbf{m}} \otimes \tilde{\mathbf{n}} + \frac{\lambda}{\lambda + 2\mu - V \sin^2\phi}\, \tilde{\mathbf{n}} \otimes \tilde{\mathbf{m}}$$
$$+ \frac{V \cos\phi \sin\phi}{\mu - V \sin^2\phi}\, (\tilde{\mathbf{m}} \otimes \tilde{\mathbf{m}} + \boldsymbol{\ell} \otimes \boldsymbol{\ell}) + \frac{V \cos\phi \sin\phi}{\lambda + 2\mu - V \sin^2\phi}\, \tilde{\mathbf{n}} \otimes \tilde{\mathbf{n}}. \tag{3.75}$$

Then from (3.69) and (3.70),

$$- \mathbf{Q}(\phi) + \mathbf{R}(\phi)\mathbf{T}(\phi)^{-1}\mathbf{R}(\phi)^T$$
$$= -(\lambda + \mu)\, \tilde{\mathbf{m}} \otimes \tilde{\mathbf{m}} - (\mu - V \cos^2\phi)\, \mathbf{I}$$
$$+ \frac{\lambda^2}{\lambda + 2\mu - V \sin^2\phi}\, \tilde{\mathbf{m}} \otimes \tilde{\mathbf{m}} + \frac{\mu^2}{\mu - V \sin^2\phi}\, \tilde{\mathbf{n}} \otimes \tilde{\mathbf{n}}$$
$$+ \left(\frac{\lambda V \cos\phi \sin\phi}{\lambda + 2\mu - V \sin^2\phi} + \frac{\mu V \cos\phi \sin\phi}{\mu - V \sin^2\phi} \right) (\tilde{\mathbf{m}} \otimes \tilde{\mathbf{n}} + \tilde{\mathbf{n}} \otimes \tilde{\mathbf{m}})$$
$$+ \frac{(V \cos\phi \sin\phi)^2}{\mu - V \sin^2\phi}\, (\tilde{\mathbf{m}} \otimes \tilde{\mathbf{m}} + \boldsymbol{\ell} \otimes \boldsymbol{\ell}) + \frac{(V \cos\phi \sin\phi)^2}{\lambda + 2\mu - V \sin^2\phi}\, \tilde{\mathbf{n}} \otimes \tilde{\mathbf{n}}. \tag{3.76}$$

On the other hand, from (3.16) we have

$$\tilde{\mathbf{m}} \otimes \tilde{\mathbf{m}} = \mathbf{m} \otimes \mathbf{m}\, \cos^2\phi + (\mathbf{m} \otimes \mathbf{n} + \mathbf{n} \otimes \mathbf{m}) \cos\phi \sin\phi + \mathbf{n} \otimes \mathbf{n}\, \sin^2\phi,$$
$$\tilde{\mathbf{m}} \otimes \tilde{\mathbf{n}} = (-\mathbf{m} \otimes \mathbf{m} + \mathbf{n} \otimes \mathbf{n}) \cos\phi \sin\phi + \mathbf{m} \otimes \mathbf{n}\, \cos^2\phi - \mathbf{n} \otimes \mathbf{m}\, \sin^2\phi,$$
$$\tilde{\mathbf{n}} \otimes \tilde{\mathbf{m}} = (-\mathbf{m} \otimes \mathbf{m} + \mathbf{n} \otimes \mathbf{n}) \cos\phi \sin\phi - \mathbf{m} \otimes \mathbf{n}\, \sin^2\phi + \mathbf{n} \otimes \mathbf{m}\, \cos^2\phi,$$
$$\tilde{\mathbf{n}} \otimes \tilde{\mathbf{n}} = \mathbf{m} \otimes \mathbf{m}\, \sin^2\phi - (\mathbf{m} \otimes \mathbf{n} + \mathbf{n} \otimes \mathbf{m}) \cos\phi \sin\phi + \mathbf{n} \otimes \mathbf{n}\, \cos^2\phi. \tag{3.77}$$

Substituting these into (3.76), we take the angular average of it over $[-\pi, \pi]$. In this process we use the following formulas of integration:

$$\frac{1}{2\pi} \int_{-\pi}^{\pi} \cos^2\phi\, d\phi = \frac{1}{2\pi} \int_{-\pi}^{\pi} \sin^2\phi\, d\phi = \frac{1}{2},$$

---

[38] For any $\mathbf{a}, \mathbf{b}, \mathbf{c}, \mathbf{d} \in \mathbb{R}^3$, $(\mathbf{a} \otimes \mathbf{b})(\mathbf{c} \otimes \mathbf{d}) = (\mathbf{c} \cdot \mathbf{b})(\mathbf{a} \otimes \mathbf{d})$.

and

$$\frac{1}{2\pi}\int_{-\pi}^{\pi}\frac{\cos^2\phi}{a-b\,\sin^2\phi}\,d\phi = \frac{1}{b}\left(1-\sqrt{\frac{a-b}{a}}\right),$$

$$\frac{1}{2\pi}\int_{-\pi}^{\pi}\frac{\sin^2\phi}{a-b\,\sin^2\phi}\,d\phi = \frac{1}{b}\left(-1+\sqrt{\frac{a}{a-b}}\right),$$

$$\frac{1}{2\pi}\int_{-\pi}^{\pi}\frac{\cos^2\phi\,\sin^2\phi}{a-b\,\sin^2\phi}\,d\phi = \frac{1}{2b^2}\left(2a-b-2\sqrt{a}\sqrt{a-b}\right),$$

$$\frac{1}{2\pi}\int_{-\pi}^{\pi}\frac{\cos^4\phi\,\sin^2\phi}{a-b\,\sin^2\phi}\,d\phi = \frac{-1}{8b^3}\left(8a^2-12ab+3b^2-8\sqrt{a}(a-b)^{\frac{3}{2}}\right),$$

$$\frac{1}{2\pi}\int_{-\pi}^{\pi}\frac{\cos^2\phi\,\sin^4\phi}{a-b\,\sin^2\phi}\,d\phi = \frac{1}{8b^3}\left(8a^2-4ab-b^2-8a^{\frac{3}{2}}\sqrt{a-b}\right), \tag{3.78}$$

where a and b are constants such that $0 < b < a$, and we use the fact that the integration of odd functions of ϕ over $[-\pi,\pi]$ gives zero. Thus, by simple but long computations, we obtain

$$\mathbf{S}_3(v) = \frac{1}{2\pi}\int_{-\pi}^{\pi}-\mathbf{Q}(\phi)+\mathbf{R}(\phi)\mathbf{T}(\phi)^{-1}\mathbf{R}(\phi)^T\,d\phi$$

$$= \frac{1}{V}\left(\sqrt{\frac{\mu}{\mu-V}}(2\mu-V)^2-\sqrt{\frac{\lambda+2\mu-V}{\lambda+2\mu}}4\mu^2\right)\mathbf{m}\otimes\mathbf{m}$$

$$+\frac{1}{V}\left(\sqrt{\frac{\lambda+2\mu}{\lambda+2\mu-V}}(2\mu-V)^2-\sqrt{\frac{\mu-V}{\mu}}4\mu^2\right)\mathbf{n}\otimes\mathbf{n}$$

$$-\sqrt{\mu(\mu-V)}\,\boldsymbol{\ell}\otimes\boldsymbol{\ell}. \tag{3.79}$$

Therefore, in terms of the orthonormal basis $\{\mathbf{m},\mathbf{n},\boldsymbol{\ell}\}$, $\mathbf{S}_3(v)$ can be written componentwise as

$$\mathbf{S}_3(v) = \begin{pmatrix} R(v) & 0 & 0 \\ 0 & \sqrt{\frac{\lambda+2\mu}{\lambda+2\mu-V}}\sqrt{\frac{\mu-V}{\mu}}R(v) & 0 \\ 0 & 0 & -\sqrt{\mu(\mu-V)} \end{pmatrix}, \tag{3.80}$$

where

$$R(v) = \frac{1}{V}\left(\sqrt{\frac{\mu}{\mu-V}}(2\mu-V)^2-\sqrt{\frac{\lambda+2\mu-V}{\lambda+2\mu}}4\mu^2\right), \quad V=\rho v^2. \tag{3.81}$$

Now we use Theorem 3.12. By (3.74) and the fact that $0 < v_R^{\text{Iso}} < v_L^{\text{Iso}}$, the $(3,3)$ component $-\sqrt{\mu(\mu-V)}$ and the multiplier $\sqrt{\frac{\lambda+2\mu}{\lambda+2\mu-V}}\sqrt{\frac{\mu-V}{\mu}}$ in the $(2,2)$ component of (3.80) do not vanish at $v=v_R^{\text{Iso}}$. Therefore, (3.54) implies that v_R^{Iso} satisfies the equation

$$R(v) = 0 \tag{3.82}$$

in the subsonic range $0 < v < \sqrt{\frac{\mu}{\rho}}$. Equation (3.82) can be written as

$$\sqrt{\frac{\mu}{\mu - V}} (2\mu - V)^2 = \sqrt{\frac{\lambda + 2\mu - V}{\lambda + 2\mu}} 4\mu^2.$$

Squaring both sides leads to

$$\frac{1}{\mu - V} (2\mu - V)^4 = \frac{\lambda + 2\mu - V}{\lambda + 2\mu} 16\mu^3,$$

and hence

$$(\lambda + 2\mu)(2\mu - V)^4 = (\mu - V)(\lambda + 2\mu - V) 16\mu^3,$$

which is a quartic equation in V. But the constant term vanishes and we get the cubic equation

$$V^3 - 8\mu V^2 + \frac{8\mu^2(3\lambda + 4\mu)}{\lambda + 2\mu} V - \frac{16\mu^3(\lambda + \mu)}{\lambda + 2\mu} = 0. \tag{3.83}$$

Proposition 3.17 *The velocity of Rayleigh waves which propagate along the surface of an isotropic elastic half-space is uniquely determined. It is the simple solution* $v = v_R^{\text{Iso}}$ *to (3.83) with* $V = \rho v^2$ *in the subsonic range* $0 < v < \sqrt{\mu/\rho}$.

Proof It is sufficient to prove that in the interval $0 < V < \mu$, equation (3.83) has only one solution. Write the left hand side of (3.83) by $\tilde{R}(V)$. Then

$$\tilde{R}(0) = \frac{-16\mu^3(\lambda + \mu)}{\lambda + 2\mu}$$

which, by (1.83), is negative and

$$\tilde{R}(\mu) = \mu^3 > 0.$$

Hence $\tilde{R}(V) = 0$ has at least one solution in the interval $0 < V < \mu$. Moreover, it is easy to see that

$$\frac{d^2}{dV^2} \tilde{R}(V) = 6V - 16\mu < 0$$

for $0 < V < \mu$, i.e., $\tilde{R}(V)$ is concave in the interval $0 < V < \mu$. Thus, the solution in the interval $0 < V < \mu$ is unique and simple. □

Next we compute the polarization vector $a_{\text{pol}}^{\text{Iso}}$ of the Rayleigh waves at the surface $\mathbf{n} \cdot \mathbf{x} = 0$. To make use of Theorem 3.16, we first compute $\mathbf{S}_1(v)$, whose integrand is the negative of (3.75). Substituting (3.77) into (3.75), we take the angular average of it over $[-\pi, \pi]$. In this process, we use the first three integration formulas in (3.78). The result is

$$-\mathbf{S}_1(v) = \frac{1}{2\pi} \int_{-\pi}^{\pi} \mathbf{T}(\phi)^{-1} \mathbf{R}(\phi)^T \, d\phi$$

$$= \frac{1}{V} \left(\sqrt{\frac{\lambda + 2\mu}{\lambda + 2\mu - V}} (2\mu - V) - \sqrt{\frac{\mu - V}{\mu}} 2\mu \right) \mathbf{m} \otimes \mathbf{n}$$

$$+ \frac{1}{V} \left(\sqrt{\frac{\lambda + 2\mu - V}{\lambda + 2\mu}} 2\mu - \sqrt{\frac{\mu}{\mu - V}} (2\mu - V) \right) \mathbf{n} \otimes \mathbf{m}.$$

Hence, in terms of the orthonormal basis $\{\mathbf{m}, \mathbf{n}, \boldsymbol{\ell}\}$, $\mathbf{S}_1(v)$ can be written componentwise as

$$\mathbf{S}_1(v) = \begin{pmatrix} 0 & -S(v) & 0 \\ \sqrt{\frac{\mu}{\mu-V}}\sqrt{\frac{\lambda+2\mu-V}{\lambda+2\mu}} \, S(v) & 0 & 0 \\ 0 & 0 & 0 \end{pmatrix}, \tag{3.84}$$

where

$$S(v) = \frac{1}{V} \left(\sqrt{\frac{\lambda + 2\mu}{\lambda + 2\mu - V}} (2\mu - V) - \sqrt{\frac{\mu - V}{\mu}} 2\mu \right). \tag{3.85}$$

Hereafter, we represent vectors componentwise in terms of the orthonormal basis $\{\mathbf{m}, \mathbf{n}, \boldsymbol{\ell}\}$, too.

Now we apply Theorem 3.16 to isotropic elasticity. Take

$$\mathbf{e}_1 = \begin{pmatrix} 0 \\ 0 \\ 1 \end{pmatrix}.$$

Then from (3.80) we get

$$\mathbf{S}_3(v_R^{\mathrm{Iso}})\mathbf{e}_1 = \begin{pmatrix} 0 \\ 0 \\ -\sqrt{\mu(\mu - V_R^{\mathrm{Iso}})} \end{pmatrix},$$

where $V_R^{\mathrm{Iso}} = \rho \, (v_R^{\mathrm{Iso}})^2$. Then by (3.74), $\mathbf{S}_3(v_R^{\mathrm{Iso}})\mathbf{e}_1 \neq \mathbf{0}$. Next, taking

$$\mathbf{e}_2 = \begin{pmatrix} 1 \\ 0 \\ 0 \end{pmatrix},$$

we have

$$(\mathbf{S}_3(v_R^{\mathrm{Iso}})\mathbf{e}_1) \times \mathbf{e}_2 = \begin{pmatrix} 0 \\ -\sqrt{\mu(\mu - V_R^{\mathrm{Iso}})} \\ 0 \end{pmatrix} \neq \mathbf{0},$$

which is the real part of $\mathbf{a}_{\text{pol}}^{\text{Iso}}$. Then by (3.84), we get

$$-\mathbf{S}_1(v_R^{\text{Iso}})\left[(\mathbf{S}_3(v_R^{\text{Iso}})\mathbf{e}_1)\times\mathbf{e}_2\right]=\begin{pmatrix}-S(v_R^{\text{Iso}})\sqrt{\mu(\mu-V_R^{\text{Iso}})}\\0\\0\end{pmatrix},$$

which is the imaginary part of $\mathbf{a}_{\text{pol}}^{\text{Iso}}$. Thus, we obtain

$$\mathbf{a}_{\text{pol}}^{\text{Iso}}=-\sqrt{\mu(\mu-V_R^{\text{Iso}})}\begin{pmatrix}\sqrt{-1}\,S(v_R^{\text{Iso}})\\1\\0\end{pmatrix}. \tag{3.86}$$

Therefore, from (3.61) and (3.86) the real displacement field of the Rayleigh waves at the surface $\mathbf{n}\cdot\mathbf{x}=0$ can be written as

$$-\sqrt{\mu(\mu-V_R^{\text{Iso}})}$$

$$\times\left\{\begin{pmatrix}0\\1\\0\end{pmatrix}\cos k(\mathbf{m}\cdot\mathbf{x}-v_R^{\text{Iso}}t)+\begin{pmatrix}S(v_R^{\text{Iso}})\\0\\0\end{pmatrix}\sin k(\mathbf{m}\cdot\mathbf{x}-v_R^{\text{Iso}}t)\right\}. \tag{3.87}$$

This is a componentwise representation relative to the orthonormal basis $\{\mathbf{m},\mathbf{n},\boldsymbol{\ell}\}$. Recalling that $\mathbf{m}$ is the propagation direction and $\mathbf{n}$ is the unit outward normal to the surface, we obtain

Proposition 3.18 *The longitudinal component and the normal component of the displacements of the Rayleigh waves at the surface of an isotropic elastic half-space have a phase shift of 90 degrees. The polarization ratio, i.e., the ratio of the maximum longitudinal component to the maximum normal component, is*

$$S(v_R^{\text{Iso}})=\frac{2\sqrt{\mu(\mu-V_R^{\text{Iso}})}}{2\mu-V_R^{\text{Iso}}}=\frac{1}{2\mu}\sqrt{\frac{\lambda+2\mu}{\lambda+2\mu-V_R^{\text{Iso}}}}\,(2\mu-V_R^{\text{Iso}}), \tag{3.88}$$

where $V_R^{\text{Iso}}=\rho\left(v_R^{\text{Iso}}\right)^2$ and v_R^{Iso} is the velocity of Rayleigh waves in Proposition 3.17.

Note The equalities in (3.88) follow from (3.82) and (3.85).

3.5 Rayleigh Waves in Weakly Anisotropic Elastic Media

In this section we assume that the elasticity tensor $\mathbf{C}=\left(C_{ijkl}\right)_{i,j,k,l=1,2,3}$ has the form

$$\mathbf{C}=\mathbf{C}^{\text{Iso}}+\mathbf{A}, \tag{3.89}$$

where $\mathbf{C}^{\text{Iso}}$ is the isotropic part of $\mathbf{C}$,

$$\mathbf{C}^{\text{Iso}}=\left(C_{ijkl}^{\text{Iso}}\right)_{i,j,k,l=1,2,3},\qquad C_{ijkl}^{\text{Iso}}=\lambda\,\delta_{ij}\delta_{kl}+\mu\left(\delta_{ik}\delta_{jl}+\delta_{il}\delta_{kj}\right) \tag{3.90}$$

with Lamé constants λ and μ, and $\mathbf{A}$ is the perturbative part of $\mathbf{C}$

$$\mathbf{A}=\left(a_{ijkl}\right)_{i,j,k,l=1,2,3}.$$

From the symmetry conditions (1.4) and (1.6) of the elasticity tensor $\mathbf{C}$ it follows that

$$a_{ijkl} = a_{jikl} = a_{klij}, \qquad i, j, k, l = 1, 2, 3,$$

but we do not assume any other symmetry for $\mathbf{A}$. Hence the perturbative part $\mathbf{A}$ has 21 independent components and we write them in the Voigt notations as

$$\mathbf{A} = (a_{rs}) = \begin{bmatrix} a_{11} & a_{12} & a_{13} & a_{14} & a_{15} & a_{16} \\ & a_{22} & a_{23} & a_{24} & a_{25} & a_{26} \\ & & a_{33} & a_{34} & a_{35} & a_{36} \\ & & & a_{44} & a_{45} & a_{46} \\ & & & & a_{55} & a_{56} \\ \text{Sym.} & & & & & a_{66} \end{bmatrix}. \tag{3.91}$$

Let $\mathbf{m}$ and $\mathbf{n}$ be orthogonal unit vectors in $\mathbb{R}^3$. Assuming that the deviation from an isotropic state caused by the perturbative part $\mathbf{A}$ is small, we consider Rayleigh waves in the half-space $\mathbf{n} \cdot \mathbf{x} \leq 0$ which propagate along the surface $\mathbf{n} \cdot \mathbf{x} = 0$ in the direction of $\mathbf{m}$ with the phase velocity v_R. In this section we investigate the perturbation of the phase velocity v_R of Rayleigh waves, i.e., the shift in v_R from its comparative isotropic value v_R^{Iso}, caused by the perturbative part $\mathbf{A}$. We present a velocity formula which is correct to within terms linear in the components of $\mathbf{A} = (a_{rs})$. Hereafter, without loss of generality, we take

$$\mathbf{m} = (0, 1, 0), \qquad \mathbf{n} = (0, 0, 1). \tag{3.92}$$

Theorem 3.19 *In a weakly anisotropic elastic medium whose elasticity tensor $\mathbf{C}$ is given by (3.89), (3.90) and (3.91), the phase velocity of Rayleigh waves which propagate along the surface of the half-space $x_3 \leq 0$ in the direction of the 2-axis can be written, to within terms linear in the perturbative part $\mathbf{A} = (a_{rs})$, as*

$$v_R = v_R^{\text{Iso}} - \frac{1}{2\rho\, v_R^{\text{Iso}}}$$

$$\times \left[\gamma_{22}\big(v_R^{\text{Iso}}\big)\, a_{22} + \gamma_{23}\big(v_R^{\text{Iso}}\big)\, a_{23} + \gamma_{33}\big(v_R^{\text{Iso}}\big)\, a_{33} + \gamma_{44}\big(v_R^{\text{Iso}}\big)\, a_{44} \right], \tag{3.93}$$

where

$$\gamma_{22}(v) = \frac{(\lambda + 2\mu)\left[-8\mu^2(\lambda + \mu) + 2\mu(5\lambda + 6\mu)V - (2\lambda + 3\mu)V^2\right]}{D(v)},$$

$$\gamma_{23}(v) = \frac{4\lambda(\mu - V)\left[4\mu(\lambda + \mu) - (\lambda + 2\mu)V\right]}{D(v)},$$

$$\gamma_{33}(v) = \frac{(\lambda + 2\mu - V)\left[-8\mu^2(\lambda + \mu) + 2\mu(5\lambda + 6\mu)V - (2\lambda + 3\mu)V^2\right]}{D(v)}$$

$$= \left(1 - \frac{V}{\lambda + 2\mu}\right)\gamma_{22}(v),$$

$$\gamma_{44}(v) = -\frac{8\mu(\lambda + 2\mu - V)\left[2\mu(\lambda + \mu) - (\lambda + 2\mu)V\right]}{D(v)},$$

$$D(v) = (\lambda + \mu)\left[8\mu^2(3\lambda + 4\mu) - 16\mu(\lambda + 2\mu)V + 3(\lambda + 2\mu)V^2\right],$$

$$V = \rho\, v^2 \tag{3.94}$$

and v_R^{Iso} is the velocity of Rayleigh waves in the comparative isotropic medium with $\mathbf{C} = \mathbf{C}^{\mathrm{Iso}}$, i.e., $V_R^{\mathrm{Iso}} = \rho \left(v_R^{\mathrm{Iso}} \right)^2$ is the unique solution to the cubic equation (3.83) in the range $0 < V < \mu$.

Remark 3.20 Only four components a_{22}, a_{23}, a_{33} and a_{44} of the perturbative part $\mathbf{A}$ can affect the first-order perturbation of the phase velocity v_R.

To prove the theorem, we make use of Theorem 3.12. Since we are concerned with the terms in v_R up to those linear in the perturbative part $\mathbf{A}$, we first express $S_3(v)$ in (3.39), to within terms linear in $\mathbf{A}$. Then we obtain an approximate secular equation for v_R, which contains terms up to those linear in $\mathbf{A}$. Hereafter we denote the matrices $\mathbf{Q}(\phi)$, $\mathbf{R}(\phi)$ and $\mathbf{T}(\phi)$ for an isotropic state given in (3.69), (3.70) and (3.71) by $\mathbf{Q}^{\mathrm{Iso}}(\phi)$, $\mathbf{R}^{\mathrm{Iso}}(\phi)$ and $\mathbf{T}^{\mathrm{Iso}}(\phi)$, respectively, and denote by $\mathbf{Q}(\phi), \mathbf{R}(\phi), \mathbf{T}(\phi)$ those of the material whose elasticity tensor is given by (3.89). Then from (3.17) and (3.89) we have

$$\mathbf{Q}(\phi) = \mathbf{Q}^{\mathrm{Iso}}(\phi) + \mathbf{Q}^{\mathrm{Ptb}}(\phi), \qquad \mathbf{R}(\phi) = \mathbf{R}^{\mathrm{Iso}}(\phi) + \mathbf{R}^{\mathrm{Ptb}}(\phi),$$
$$\mathbf{T}(\phi) = \mathbf{T}^{\mathrm{Iso}}(\phi) + \mathbf{T}^{\mathrm{Ptb}}(\phi),$$

where

$$\mathbf{Q}^{\mathrm{Ptb}}(\phi) = \left(\sum_{j,l=1}^{3} a_{ijkl}\, \tilde{m}_j \tilde{m}_l \right)_{i\downarrow k \to 1,2,3},$$

$$\mathbf{R}^{\mathrm{Ptb}}(\phi) = \left(\sum_{j,l=1}^{3} a_{ijkl}\, \tilde{m}_j \tilde{n}_l \right)_{i\downarrow k \to 1,2,3},$$

$$\mathbf{T}^{\mathrm{Ptb}}(\phi) = \left(\sum_{j,l=1}^{3} a_{ijkl}\, \tilde{n}_j \tilde{n}_l \right)_{i\downarrow k \to 1,2,3}. \tag{3.95}$$

Then

$$\mathbf{T}(\phi)^{-1} = \left(\mathbf{T}^{\mathrm{Iso}}(\phi) + \mathbf{T}^{\mathrm{Ptb}}(\phi) \right)^{-1} = \left(\mathbf{T}^{\mathrm{Iso}}(\phi) \left(\mathbf{I} + \mathbf{T}^{\mathrm{Iso}}(\phi)^{-1} \mathbf{T}^{\mathrm{Ptb}}(\phi) \right) \right)^{-1}$$
$$= \left(\mathbf{I} + \mathbf{T}^{\mathrm{Iso}}(\phi)^{-1} \mathbf{T}^{\mathrm{Ptb}}(\phi) \right)^{-1} \mathbf{T}^{\mathrm{Iso}}(\phi)^{-1}$$
$$\approx \left(\mathbf{I} - \mathbf{T}^{\mathrm{Iso}}(\phi)^{-1} \mathbf{T}^{\mathrm{Ptb}}(\phi) \right) \mathbf{T}^{\mathrm{Iso}}(\phi)^{-1}.$$

Here and hereafter we use the notation $\approx$ to indicate that we are retaining terms up to those linear in the perturbative part $\mathbf{A} = (a_{rs})$ and are neglecting the higher order terms. Hence

$$\mathbf{T}(\phi)^{-1} \approx \mathbf{T}^{\mathrm{Iso}}(\phi)^{-1} - \mathbf{T}^{\mathrm{Iso}}(\phi)^{-1} \mathbf{T}^{\mathrm{Ptb}}(\phi) \mathbf{T}^{\mathrm{Iso}}(\phi)^{-1}.$$

Then the integrand of $S_3(v)$ becomes

$$- \mathbf{Q}(\phi) + \mathbf{R}(\phi)\mathbf{T}(\phi)^{-1}\mathbf{R}^T(\phi)$$

$$\approx -\mathbf{Q}^{\mathrm{Iso}}(\phi) - \mathbf{Q}^{\mathrm{Ptb}}(\phi) + \left(\mathbf{R}^{\mathrm{Iso}}(\phi) + \mathbf{R}^{\mathrm{Ptb}}(\phi)\right)$$

$$\times \left(\mathbf{T}^{\mathrm{Iso}}(\phi)^{-1} - \mathbf{T}^{\mathrm{Iso}}(\phi)^{-1} \, \mathbf{T}^{\mathrm{Ptb}}(\phi) \, \mathbf{T}^{\mathrm{Iso}}(\phi)^{-1}\right) \left(\mathbf{R}^{\mathrm{Iso}}(\phi)^T + \mathbf{R}^{\mathrm{Ptb}}(\phi)^T\right)$$

$$\approx -\mathbf{Q}^{\mathrm{Iso}}(\phi) + \mathbf{R}^{\mathrm{Iso}}(\phi)\mathbf{T}^{\mathrm{Iso}}(\phi)^{-1}\mathbf{R}^{\mathrm{Iso}}(\phi)^T$$

$$- \mathbf{Q}^{\mathrm{Ptb}}(\phi) + \mathbf{R}^{\mathrm{Ptb}}(\phi)\mathbf{T}^{\mathrm{Iso}}(\phi)^{-1}\mathbf{R}^{\mathrm{Iso}}(\phi)^T + \mathbf{R}^{\mathrm{Iso}}(\phi)\mathbf{T}^{\mathrm{Iso}}(\phi)^{-1}\mathbf{R}^{\mathrm{Ptb}}(\phi)^T$$

$$- \mathbf{R}^{\mathrm{Iso}}(\phi)\mathbf{T}^{\mathrm{Iso}}(\phi)^{-1}\mathbf{T}^{\mathrm{Ptb}}(\phi)\mathbf{T}^{\mathrm{Iso}}(\phi)^{-1}\mathbf{R}^{\mathrm{Iso}}(\phi)^T.$$

Therefore, we obtain

$$\mathbf{S}_3(v) = \frac{1}{2\pi}\int_{-\pi}^{\pi}\left(-\mathbf{Q}(\phi) + \mathbf{R}(\phi)\mathbf{T}^{-1}(\phi)\mathbf{R}^T(\phi)\right)\, d\phi \approx \mathbf{S}_3^{\mathrm{Iso}}(v) + \mathbf{S}_3^{\mathrm{Ptb}}(v);$$

here

$$\mathbf{S}_3^{\mathrm{Iso}}(v) = \frac{1}{2\pi}\int_{-\pi}^{\pi}\left(-\mathbf{Q}^{\mathrm{Iso}}(\phi) + \mathbf{R}^{\mathrm{Iso}}(\phi)\mathbf{T}^{\mathrm{Iso}}(\phi)^{-1}(\phi)\mathbf{R}^{\mathrm{Iso}}(\phi)^T\right)\, d\phi$$

is of zeroth order in the perturbative part $\mathbf{A}$, which is equal to (3.79), and

$$\mathbf{S}_3^{\mathrm{Ptb}}(v) = \frac{1}{2\pi}\int_{-\pi}^{\pi}\left(-\mathbf{Q}^{\mathrm{Ptb}}(\phi) + \mathbf{R}^{\mathrm{Ptb}}(\phi)\mathbf{T}^{\mathrm{Iso}}(\phi)^{-1}\mathbf{R}^{\mathrm{Iso}}(\phi)^T\right.$$

$$+ \mathbf{R}^{\mathrm{Iso}}(\phi)\mathbf{T}^{\mathrm{Iso}}(\phi)^{-1}\mathbf{R}^{\mathrm{Ptb}}(\phi)^T$$

$$\left. - \mathbf{R}^{\mathrm{Iso}}(\phi)\mathbf{T}^{\mathrm{Iso}}(\phi)^{-1}\mathbf{T}^{\mathrm{Ptb}}(\phi)\mathbf{T}^{\mathrm{Iso}}(\phi)^{-1}\mathbf{R}^{\mathrm{Iso}}(\phi)^T\right)d\phi,$$

$$(3.96)$$

which is of first order in the perturbative part $\mathbf{A}$.

Now we prove

Lemma 3.21 *Let $s_{ij}(v)$ $(i, j = 1, 2, 3)$ be the (i, j) component of the matrix $\mathbf{S}_3^{\mathrm{Ptb}}(v)$. Then an approximate secular equation for v_R is*

$$\Delta(v) = 0,$$

where

$$\Delta(v) = R(v) + s_{22}(v), \qquad (3.97)$$

and $R(v)$ is given by (3.81).

Proof From (3.79) and (3.92) it follows that

$$\mathbf{S}_3^{\mathrm{Iso}}(v) = \begin{pmatrix} -\sqrt{\mu(\mu - V)} & 0 & 0 \\ 0 & R(v) & 0 \\ 0 & 0 & \sqrt{\frac{\lambda+2\mu}{\lambda+2\mu-V}}\sqrt{\frac{\mu-V}{\mu}}\,R(v) \end{pmatrix}.$$

Then

$$\mathbf{S}_3(v) \approx \mathbf{S}_3^{\text{Iso}}(v) + \mathbf{S}_3^{\text{Ptb}}(v) \tag{3.98}$$

$$= \begin{pmatrix} -\sqrt{\mu(\mu - V)} + s_{11}(v) & s_{12}(v) & s_{13}(v) \\ s_{12}(v) & R(v) + s_{22}(v) & s_{23}(v) \\ s_{13}(v) & s_{23}(v) & \sqrt{\frac{\lambda+2\mu}{\lambda+2\mu-V}}\sqrt{\frac{\mu-V}{\mu}} R(v) + s_{33}(v) \end{pmatrix}.$$

By Theorem 3.12, the $(3, 3)$ minor of the preceding matrix, i.e., the determinant of the submatrix formed by striking out the third row and third column, must vanish at $v = v_R$. Since $s_{ij}(v)$ $(i, j = 1, 2, 3)$ are linear functions of the perturbative part $\mathbf{A}$, the $(3, 3)$ minor becomes

$$\det \begin{pmatrix} -\sqrt{\mu(\mu - V)} + s_{11}(v) & s_{12}(v) \\ s_{12}(v) & R(v) + s_{22}(v) \end{pmatrix}$$

$$= \left(-\sqrt{\mu(\mu - V)} + s_{11}(v) \right) (R(v) + s_{22}(v)) - s_{12}(v)^2$$

$$\approx -\sqrt{\mu(\mu - V)} (R(v) + s_{22}(v)) + s_{11}(v) R(v). \tag{3.99}$$

Also, recalling that

$$R(v_R^{\text{Iso}}) = 0,$$

we have

$$s_{11}(v) R(v) \approx 0$$

at $v = v_R$. Since v_R^{Iso} is less than $\sqrt{\frac{\mu}{\rho}}$ and v_R is close to v_R^{Iso}, we get

$$\sqrt{\mu(\mu - V)} \neq 0$$

at $v = v_R$. Hence (3.99) and Theorem 3.12 imply[39] that

$$R(v) + s_{22}(v) = \Delta(v) = 0$$

at $v = v_R$. This proves the lemma.   □

Lemma 3.22

$$v_R \approx v_R^{\text{Iso}} - \sum \frac{\left. \dfrac{\partial s_{22}(v)}{\partial a_{rs}} \right|_{\mathbf{A}=(a_{rs})=0, v=v_R^{\text{Iso}}}}{\left. \dfrac{\partial R}{\partial v} \right|_{v=v_R^{\text{Iso}}}} \times a_{rs}, \tag{3.100}$$

where the summation on the right hand side is taken for the indices of the 21 independent components in (3.91).

---

[39] In the proof, we have used the vanishing of the $(3, 3)$ minor of the matrix $\mathbf{S}_3^{\text{Iso}}(v) + \mathbf{S}_3^{\text{Ptb}}(v)$ as an approximate secular equation. The vanishing of the other minors may give other approximate secular equations. However, by Theorem 3.14, the Rayleigh wave is unique if it exists. Hence there is no need to consider other secular equations.

Proof From the Taylor expansion of v_R around $\mathbf{A} = (a_{rs}) = \mathbf{0}$, we get

$$v_R \approx v_R^{\text{Iso}} + \sum \frac{\partial v_R}{\partial a_{rs}}\bigg|_{(a_{rs})=0} a_{rs}. \tag{3.101}$$

Since we have

$$\Delta(v_R) = 0,$$

it follows from the implicit function theorem that

$$\frac{\partial v_R}{\partial a_{rs}}\bigg|_{(a_{rs})=0} = -\frac{\partial \Delta}{\partial a_{rs}}\bigg|_{(a_{rs})=0,\ v=v_R^{\text{Iso}}} \left(\frac{\partial \Delta}{\partial v}\bigg|_{(a_{rs})=0,\ v=v_R^{\text{Iso}}}\right)^{-1}. \tag{3.102}$$

From (3.97) we get

$$\frac{\partial \Delta}{\partial a_{rs}} = \frac{\partial s_{22}(v)}{\partial a_{rs}} \quad \text{and} \quad \frac{\partial \Delta}{\partial v} = \frac{\partial R(v)}{\partial v} + \frac{\partial s_{22}(v)}{\partial v}. \tag{3.103}$$

Since $s_{22}(v)$ is a linear function of $\mathbf{A} = (a_{rs})$, it follows that

$$\frac{\partial s_{22}(v)}{\partial v}\bigg|_{(a_{rs})=0} = 0.$$

Hence we have

$$\frac{\partial \Delta}{\partial v}\bigg|_{(a_{rs})=0,\ v=v_R^{\text{Iso}}} = \frac{\partial R(v)}{\partial v}\bigg|_{v=v_R^{\text{Iso}}}. \tag{3.104}$$

Therefore, from (3.101)–(3.104) we obtain (3.100).[40] $\square$

The next lemma is useful in the computations of the perturbation formula.

Lemma 3.23 *The effect of the perturbative part* $\mathbf{A} = (a_{rs})$ *on* $s_{22}(v)$, *i.e., on the* $(2, 2)$ *component of the matrix* $\mathbf{S}_3^{\text{Ptb}}(v)$, *comes from* a_{22}, a_{23}, a_{33} *and* a_{44}.

Proof First we prove that the effect of the perturbative part $\mathbf{A}$ on the $(2, 2)$ component of the integrand of $\mathbf{S}_3^{\text{Ptb}}(v)$ (see (3.96)), i.e., on the $(2, 2)$ component of

$$-\mathbf{Q}^{\text{Ptb}}(\phi) + \mathbf{R}^{\text{Ptb}}(\phi)\mathbf{T}^{\text{Iso}}(\phi)^{-1}\mathbf{R}^{\text{Iso}}(\phi)^T + \mathbf{R}^{\text{Iso}}(\phi)\mathbf{T}^{\text{Iso}}(\phi)^{-1}\mathbf{R}^{\text{Ptb}}(\phi)^T$$
$$-\mathbf{R}^{\text{Iso}}(\phi)\mathbf{T}^{\text{Iso}}(\phi)^{-1}\mathbf{T}^{\text{Ptb}}(\phi)\mathbf{T}^{\text{Iso}}(\phi)^{-1}\mathbf{R}^{\text{Iso}}(\phi)^T, \tag{3.105}$$

comes from $a_{22}, a_{23}, a_{33}, a_{44}, a_{24}$ and a_{34}.

From (3.16) and (3.92) we have

$$\tilde{\mathbf{m}} = (0, \tilde{m}_2, \tilde{m}_3) = (0, \cos\phi, \sin\phi),$$
$$\tilde{\mathbf{n}} = (0, \tilde{n}_2, \tilde{n}_3) = (0, -\sin\phi, \cos\phi). \tag{3.106}$$

---

[40]By Proposition 3.17, the unique solution to the cubic equation (3.83) in the range $0 < V < \mu$ is simple. Hence (3.104), i.e., the denominator of (3.102), does not vanish.

Then from (3.91) and (3.95) we get

$$
\mathbf{Q}^{\mathrm{Ptb}}(\phi) = \begin{bmatrix} a_{66}\cos^2\phi + 2a_{56}\cos\phi\sin\phi + a_{55}\sin^2\phi & a_{26}\cos^2\phi + (a_{46}+a_{25})\cos\phi\sin\phi + a_{45}\sin^2\phi \\ & a_{22}\cos^2\phi + 2a_{24}\cos\phi\sin\phi + a_{44}\sin^2\phi \\ \mathrm{Sym.} & \end{bmatrix}
$$

$$
\begin{matrix} a_{46}\cos^2\phi + (a_{36}+a_{45})\cos\phi\sin\phi + a_{35}\sin^2\phi \\ a_{24}\cos^2\phi + (a_{23}+a_{44})\cos\phi\sin\phi + a_{34}\sin^2\phi \\ a_{44}\cos^2\phi + 2a_{34}\cos\phi\sin\phi + a_{33}\sin^2\phi \end{matrix} \Bigg], \tag{3.107}
$$

$$
\mathbf{R}^{\mathrm{Ptb}}(\phi) = \begin{bmatrix} (a_{55}-a_{66})\cos\phi\sin\phi + a_{56}(\cos^2\phi-\sin^2\phi) & (a_{45}-a_{26})\cos\phi\sin\phi + a_{46}\cos^2\phi - a_{25}\sin^2\phi \\ (a_{45}-a_{26})\cos\phi\sin\phi + a_{25}\cos^2\phi - a_{46}\sin^2\phi & (a_{44}-a_{22})\cos\phi\sin\phi + a_{24}(\cos^2\phi-\sin^2\phi) \\ (a_{35}-a_{46})\cos\phi\sin\phi + a_{45}\cos^2\phi - a_{36}\sin^2\phi & (a_{34}-a_{24})\cos\phi\sin\phi + a_{44}\cos^2\phi - a_{23}\sin^2\phi \end{bmatrix}
$$

$$
\begin{matrix} (a_{35}-a_{46})\cos\phi\sin\phi + a_{36}\cos^2\phi - a_{45}\sin^2\phi \\ (a_{34}-a_{24})\cos\phi\sin\phi + a_{23}\cos^2\phi - a_{44}\sin^2\phi \\ (a_{33}-a_{44})\cos\phi\sin\phi + a_{34}(\cos^2\phi-\sin^2\phi) \end{matrix} \Bigg], \tag{3.108}
$$

$$
\mathbf{T}^{\mathrm{Ptb}}(\phi) = \begin{bmatrix} a_{55}\cos^2\phi - 2a_{56}\cos\phi\sin\phi + a_{66}\sin^2\phi & a_{45}\cos^2\phi - (a_{46}+a_{25})\cos\phi\sin\phi + a_{26}\sin^2\phi \\ & a_{44}\cos^2\phi - 2a_{24}\cos\phi\sin\phi + a_{22}\sin^2\phi \\ \mathrm{Sym.} & \end{bmatrix}
$$

$$
\begin{matrix} a_{35}\cos^2\phi - (a_{36}+a_{45})\cos\phi\sin\phi + a_{46}\sin^2\phi \\ a_{34}\cos^2\phi - (a_{23}+a_{44})\cos\phi\sin\phi + a_{24}\sin^2\phi \\ a_{33}\cos^2\phi - 2a_{34}\cos\phi\sin\phi + a_{44}\sin^2\phi \end{matrix} \Bigg]. \tag{3.109}
$$

From the above we see that the (i, j) components $(2 \le i, j \le 3)$ of the matrices $\mathbf{Q}^{\mathrm{Ptb}}(\phi)$, $\mathbf{R}^{\mathrm{Ptb}}(\phi)$, $\mathbf{T}^{\mathrm{Ptb}}(\phi)$ depend only on $a_{22}, a_{23}, a_{33}, a_{44}, a_{24}, a_{34}$, whereas the other components depend only on $a_{25}, a_{26}, a_{35}, a_{36}, a_{45}, a_{46}, a_{55}, a_{56}, a_{66}$. Let us write these matrices symbolically as

$$
\mathbf{M}^{\mathrm{Ptb}} = \begin{bmatrix} \triangle & \triangle & \triangle \\ \triangle & \square & \square \\ \triangle & \square & \square \end{bmatrix}, \tag{3.110}
$$

where an entry of '$\triangle$' denotes a component the effect of the perturbative part $\mathbf{A}$ on which comes from $a_{25}, a_{26}, a_{35}, a_{36}, a_{45}, a_{46}, a_{55}, a_{56}, a_{66}$, and that of '$\square$' denotes a component the effect of $\mathbf{A}$ on which comes from $a_{22}, a_{23}, a_{33}, a_{44}, a_{24}, a_{34}$.

On the other hand, from (3.70), (3.73) and (3.106) we have

$$
\mathbf{T}^{\mathrm{Iso}}(\phi)^{-1} = \begin{bmatrix} \dfrac{1}{\mu - V\sin^2\phi} & 0 \\[2mm] & \dfrac{\cos^2\phi}{\mu - V\sin^2\phi} + \dfrac{\sin^2\phi}{\lambda + 2\mu - V\sin^2\phi} \\[2mm] \mathrm{Sym.} & \end{bmatrix}
$$

$$
\begin{matrix} 0 \\[2mm] \left(\dfrac{1}{\mu - V\sin^2\phi} - \dfrac{1}{\lambda + 2\mu - V\sin^2\phi}\right)\cos\phi\sin\phi \\[2mm] \dfrac{\sin^2\phi}{\mu - V\sin^2\phi} + \dfrac{\cos^2\phi}{\lambda + 2\mu - V\sin^2\phi} \end{matrix} \Bigg] \tag{3.111}
$$

and

$$\mathbf{R}^{\text{Iso}}(\phi) = \begin{bmatrix} V\cos\phi\sin\phi & 0 & 0 \\ 0 & (V-\lambda-\mu)\cos\phi\sin\phi & \lambda\cos^2\phi-\mu\sin^2\phi \\ 0 & \mu\cos^2\phi-\lambda\sin^2\phi & (V+\lambda+\mu)\cos\phi\sin\phi \end{bmatrix}. \quad (3.112)$$

The $(i,1)$ and $(1,i)$ components $(i=2,3)$ of the matrices $\mathbf{T}^{\text{Iso}}(\phi)^{-1}$, $\mathbf{R}^{\text{Iso}}(\phi)$ vanish, so do those of the matrices $\mathbf{T}^{\text{Iso}}(\phi)^{-1}\mathbf{R}^{\text{Iso}}(\phi)^T$, $\mathbf{R}^{\text{Iso}}(\phi)\mathbf{T}^{\text{Iso}}(\phi)^{-1}$. We write these matrices symbolically as

$$\mathbf{M}^{\text{Iso}} = \begin{bmatrix} * & 0 & 0 \\ 0 & * & * \\ 0 & * & * \end{bmatrix}, \quad (3.113)$$

where '$*$' denotes a possibly non-zero component.

Now let us evaluate the products of matrices in the integrand (3.105). We see from (3.110) and (3.113) that

$$\mathbf{M}^{\text{Ptb}}\mathbf{M}^{\text{Iso}} = \begin{bmatrix} \triangle & \triangle & \triangle \\ \triangle & \square & \square \\ \triangle & \square & \square \end{bmatrix},$$

and

$$\mathbf{M}^{\text{Iso}}\mathbf{M}^{\text{Ptb}} = \begin{bmatrix} \triangle & \triangle & \triangle \\ \triangle & \square & \square \\ \triangle & \square & \square \end{bmatrix}.$$

From this observation, the effect of the perturbative part $\mathbf{A} = (a_{rs})$ on the $(2,2)$ component of (3.105) comes from $a_{22}, a_{23}, a_{33}, a_{44}, a_{24}$ and a_{34}.

Next we prove that the effect of the perturbative part $\mathbf{A} = (a_{rs})$ on $s_{22}(v)$, i.e., on the $(2,2)$ component of $\mathbf{S}_3^{\text{Ptb}}(v)$, comes only from a_{22}, a_{23}, a_{33} and a_{44}. Since (3.105) is to be integrated over the interval $[-\pi,\pi]$ with respect to ϕ (see (3.96)), it is sufficient to show that the coefficients of a_{24} and a_{34} in the $(2,2)$ component of (3.105) are odd functions in ϕ.

By (3.111) and (3.112), we can write

$$\mathbf{T}^{\text{Iso}}(\phi)^{-1} = \begin{bmatrix} even & 0 & 0 \\ 0 & even & odd \\ 0 & odd & even \end{bmatrix}$$

and

$$\mathbf{R}^{\text{Iso}}(\phi) = \begin{bmatrix} odd & 0 & 0 \\ 0 & odd & even \\ 0 & even & odd \end{bmatrix},$$

where an entry of '$even$' (resp. 'odd') denotes a component which is an even (resp. odd) function in ϕ. Then we get

$$\mathbf{T}^{\text{Iso}}(\phi)^{-1}\mathbf{R}^{\text{Iso}}(\phi)^T = \begin{bmatrix} odd & 0 & 0 \\ 0 & odd & even \\ 0 & even & odd \end{bmatrix} \quad (3.114)$$

and

$$\mathbf{R}^{\mathrm{Iso}}(\phi)\mathbf{T}^{\mathrm{Iso}}(\phi)^{-1} = \begin{bmatrix} odd & 0 & 0 \\ 0 & odd & even \\ 0 & even & odd \end{bmatrix}. \tag{3.115}$$

On the other hand, looking at the coefficients of a_{24} and a_{34} in (3.107), (3.108) and (3.109) carefully, we can write

$$\mathbf{Q}^{\mathrm{Ptb}}(\phi) = \begin{bmatrix} \times & \times & \times \\ \times & a_{24}\text{-}odd & a_{24}, a_{34}\text{-}even \\ \times & a_{24}, a_{34}\text{-}even & a_{34}\text{-}odd \end{bmatrix}, \tag{3.116}$$

$$\mathbf{R}^{\mathrm{Ptb}}(\phi) = \begin{bmatrix} \times & \times & \times \\ \times & a_{24}\text{-}even & a_{24}, a_{34}\text{-}odd \\ \times & a_{24}, a_{34}\text{-}odd & a_{34}\text{-}even \end{bmatrix}, \tag{3.117}$$

$$\mathbf{T}^{\mathrm{Ptb}}(\phi) = \begin{bmatrix} \times & \times & \times \\ \times & a_{24}\text{-}odd & a_{24}, a_{34}\text{-}even \\ \times & a_{24}, a_{34}\text{-}even & a_{34}\text{-}odd \end{bmatrix}, \tag{3.118}$$

where the label 'a_{ij}-odd' (resp. 'a_{ij}-even') signifies that the component in question has its a_{ij} coefficient being an odd (resp. even) function in ϕ, and '$\times$' denotes a component which does not depend on a_{24} and a_{34}. Thus, we obtain from (3.114) and (3.117),

$$\mathbf{R}^{\mathrm{Ptb}}(\phi)\mathbf{T}^{\mathrm{Iso}}(\phi)^{-1}\mathbf{R}^{\mathrm{Iso}}(\phi)^{T} = \begin{bmatrix} \times & \times & \times \\ \times & a_{24}, a_{34}\text{-}odd & a_{24}, a_{34}\text{-}even \\ \times & a_{24}, a_{34}\text{-}even & a_{24}, a_{34}\text{-}odd \end{bmatrix}, \tag{3.119}$$

and then,

$$\mathbf{R}^{\mathrm{Iso}}(\phi)\mathbf{T}^{\mathrm{Iso}}(\phi)^{-1}\mathbf{R}^{\mathrm{Ptb}}(\phi)^{T} = \begin{bmatrix} \times & \times & \times \\ \times & a_{24}, a_{34}\text{-}odd & a_{24}, a_{34}\text{-}even \\ \times & a_{24}, a_{34}\text{-}even & a_{24}, a_{34}\text{-}odd \end{bmatrix}, \tag{3.120}$$

and from (3.114), (3.115) and (3.118),

$$\mathbf{R}^{\mathrm{Iso}}(\phi)\mathbf{T}^{\mathrm{Iso}}(\phi)^{-1}\mathbf{T}^{\mathrm{Ptb}}(\phi)\mathbf{T}^{\mathrm{Iso}}(\phi)^{-1}\mathbf{R}^{\mathrm{Iso}}(\phi)^{T}$$

$$= \begin{bmatrix} \times & \times & \times \\ \times & a_{24}, a_{34}\text{-}odd & a_{24}, a_{34}\text{-}even \\ \times & a_{24}, a_{34}\text{-}even & a_{24}, a_{34}\text{-}odd \end{bmatrix}. \tag{3.121}$$

Then we can observe that the coefficients of a_{24} and a_{34} in the $(2, 2)$ components of (3.116), (3.119), (3.120) and (3.121) are odd functions in ϕ, and so are the coefficients of a_{24} and a_{34} in the $(2, 2)$ component of the integrand (3.105). Therefore, the terms linear in a_{24} and a_{34} included in the $(2, 2)$ component of $\mathbf{S}_3^{\mathrm{Ptb}}(v)$ vanish. This completes the proof of the lemma.   □

Proof of Theorem 3.19 [41] Let us start with an orthorhombic elastic half-space $x_3 \leq 0$ whose elasticity tensor $\mathbf{C}$ is given by (1.15) and consider Rayleigh waves that propagate along its free surface. Under the assumption that $\mathbf{C}$ is expressed by (3.89), its perturbative part $\mathbf{A}$ has the following orthorhombic form:

$$\mathbf{A} = (a_{rs}) = \begin{bmatrix} a_{11} & a_{12} & a_{13} & 0 & 0 & 0 \\ & a_{22} & a_{23} & 0 & 0 & 0 \\ & & a_{33} & 0 & 0 & 0 \\ & & & a_{44} & 0 & 0 \\ & & & & a_{55} & 0 \\ \text{Sym.} & & & & & a_{66} \end{bmatrix}.$$

Then the phase velocity of Rayleigh waves which propagate along the surface of the half-space $x_3 \leq 0$ in the direction of the 2-axis satisfies the secular equation[42]

$$R^{\text{Orth}}(v) = 0, \tag{3.122}$$

where

$$R^{\text{Orth}}(v) = C_{33}C_{44}(C_{22} - V)V^2 - (C_{44} - V)\left(C_{33}(C_{22} - V) - C_{23}^2\right)^2 \tag{3.123}$$

and $V = \rho v^2$.

The components of the perturbative part of $\mathbf{A}$ included in $R^{\text{Orth}}(v)$ are a_{22}, a_{23}, a_{33} and a_{44}. By Lemma 3.23, these are exactly the same components that affect the $(2, 2)$ component of $\mathbf{S}_3^{\text{Ptb}}(v)$. Furthermore, by Lemma 3.22, the effect of the perturbative part $\mathbf{A} = (a_{rs})$ of (3.91) on the perturbation formula comes only from these components. Hence it should also be possible to use (3.122) as a secular equation to obtain the first order perturbation formula for the velocity of Rayleigh waves which propagate along the surface of the half-space of a weakly anisotropic elastic medium whose elasticity tensor is given by (3.89), (3.90) and (3.91). In fact, by the same arguments as in Lemma 3.22, from the Taylor expansion of v_R around $\mathbf{A} = (a_{rs}) = \mathbf{0}$ we get

$$v_R \approx v_R^{\text{Iso}} + \sum \left.\frac{\partial v_R}{\partial a_{rs}}\right|_{(a_{rs})=0} a_{rs}, \tag{3.124}$$

where the summation in the second term of the right hand side is taken for the indices $(r, s) = (2, 2), (2, 3), (3, 3)$ and $(4, 4)$. Moreover, from the implicit function theorem we have

$$\left.\frac{\partial v_R}{\partial a_{ij}}\right|_{(a_{rs})=0} = -\left.\frac{\partial R^{\text{Orth}}}{\partial a_{ij}}\right|_{(a_{rs})=0,\ v=v_R^{\text{Iso}}} \left(\left.\frac{\partial R^{\text{Orth}}}{\partial v}\right|_{(a_{rs})=0,\ v=v_R^{\text{Iso}}}\right)^{-1}. \tag{3.125}$$

---

[41] Of course, it is possible to compute $s_{22}(v)$ directly by taking the angular average of the $(2, 2)$ component of (3.105) and then use Lemma 3.22 to obtain the result. But here, we take full advantage of Lemma 3.23 to derive the perturbation formula by a much simpler method.

[42] This equation is well known (see, for example, [17, 64]). It is possible to derive (3.122) from (3.47) (see Section 12.10 of [77] and Exercise 3-8).

Since $R^{\text{Orth}}(v)$ is given by (3.123) explicitly, we can calculate directly its derivatives in (3.125). For example, we get

$$\left.\frac{\partial R^{\text{Orth}}}{\partial v}\right|_{(a_{rs})=0,\; v=v_R^{\text{Iso}}} = 2\rho v_R^{\text{Iso}} \left.\frac{\partial R^{\text{Orth}}}{\partial V}\right|_{(a_{rs})=0,\; V=V_R^{\text{Iso}}=\rho(v_R^{\text{Iso}})^2}.$$

Since $C_{22} = C_{33} = \lambda + 2\mu$, $C_{44} = \mu$ and $C_{23} = \lambda$ when $(a_{rs}) = 0$, we have

$$\left.\frac{\partial R^{\text{Orth}}}{\partial V}\right|_{(a_{rs})=0,\; V=V_R^{\text{Iso}}} = \frac{d}{dV}\Big[\mu(\lambda + 2\mu)(\lambda + 2\mu - V)V^2$$
$$- (\mu - V)\big((\lambda + 2\mu)(\lambda + 2\mu - V) - \lambda^2\big)^2\Big]_{V=V_R^{\text{Iso}}}.$$

Then, by a simple computation, this becomes

$$(\lambda + \mu)\big[8\mu^2(3\lambda + 4\mu) - 16\mu(\lambda + 2\mu)V + 3(\lambda + 2\mu)V^2\big]_{V=V_R^{\text{Iso}}},$$

which is equal to $D(v)$ in (3.94). We also have

$$\left.\frac{\partial R^{\text{Orth}}}{\partial a_{22}}\right|_{(a_{rs})=0,\; v=v_R^{\text{Iso}}}$$

$$= \left.\frac{\partial R^{\text{Orth}}}{\partial C_{22}}\right|_{(a_{rs})=0,\; v=v_R^{\text{Iso}}}$$

$$= C_{33}C_{44}V^2 - 2C_{33}(C_{44} - V)\big(C_{33}(C_{22} - V) - C_{23}^2\big)\Big|_{(a_{rs})=0,\; V=V_R^{\text{Iso}}}$$

$$= (\lambda + 2\mu)\big[-8\mu^2(\lambda + \mu) + 2\mu(5\lambda + 6\mu)V - (2\lambda + 3\mu)V^2\big]_{V=V_R^{\text{Iso}}},$$

which is equal to the numerator of $\gamma_{22}(v)$ in (3.94). A similar method can be applied to get the numerators of the other $\gamma_{ij}(v)$'s in (3.94). Note that $\frac{\partial R^{\text{Orth}}}{\partial a_{44}} = \frac{\partial R^{\text{Orth}}}{\partial C_{44}}$ becomes cubic in V. In this case we use (3.83) for $V = V_R^{\text{Iso}}$ to reduce the power of V. Then we can obtain a quadratic expression in V, which can be factorized and gives the numerator of $\gamma_{44}(v)$.   □

3.6 Rayleigh Waves in Anisotropic Elasticity

Let us turn to the general anisotropic material whose elasticity tensor $\mathbf{C} = (C_{ijkl})_{i,j,k,l=1,2,3}$ has the symmetries (1.4) and (1.6) and satisfies the strong convexity condition (1.7). Then $\mathbf{C}$ has 21 independent components and is expressed under the Voigt notation as (1.12).

Let $\mathbf{m}$ and $\mathbf{n}$ be orthogonal unit vectors in $\mathbb{R}^3$. As in Section 3.3, we consider Rayleigh waves in the half-space $\mathbf{n} \cdot \mathbf{x} \leq 0$ which propagate along the surface $\mathbf{n} \cdot \mathbf{x} = 0$ in the direction of $\mathbf{m}$ with the phase velocity v_R in the subsonic range $0 < v < v_L$, where $v_L = v_L(\mathbf{m}, \mathbf{n})$ is the limiting velocity in Definition 3.1. In this section we discuss the existence of Rayleigh waves.

3.6.1 Limiting Wave Solution

In Section 3.3, we have obtained several equivalent conditions for the existence of Rayleigh waves which propagate along the surface of an elastic half-space. These conditions are given by the formulas at $v = v_R$, i.e., at the phase velocity of the Rayleigh waves. Hence they are useful in finding the secular equations for the velocity of the Rayleigh waves for isotropic and for weakly anisotropic elastic media (Section 3.4, Section 3.5).

However, for general anisotropic elasticity, the existence of Rayleigh waves is not always guaranteed, let alone obtaining their velocity analytically. Hence it is important to have certain criteria for the existence of Rayleigh waves without using their velocity.

Thus, in this section we present existence criteria that are based on the limiting velocity. Namely, we give criteria for the existence of Rayleigh waves in an elastic half-space that are expressed in terms of the behavior of $\mathbf{S}_3(v)$ and $\mathbf{Z}(v)$ in the transonic limit $v \uparrow v_L(\mathbf{m}, \mathbf{n})$, respectively, and in terms of the transonic state of the slowness section, which we shall define later.

Before presenting the existence criteria, in this subsection we establish several concepts associated with the limiting velocity $v_L(\mathbf{m}, \mathbf{n})$.

Let us recall from (2) of Lemma 3.2 and Lemma 3.3 that for velocity v such that $0 \leq v < v_L(\mathbf{m}, \mathbf{n})$ there exist surface-wave solutions to (3.3) of the form (3.27) or (3.29) or (3.31) corresponding to the three complex p_α ($\alpha = 1, 2, 3$) with Im $p_\alpha > 0$.

When $v = v_L(\mathbf{m}, \mathbf{n})$, either one, two, or three p_α (Im $p_\alpha > 0$) are found to be real. In fact, from (3.17), (3.20) and (3.26) there exists at least one $\hat{\phi}$ $\left(-\frac{\pi}{2} < \hat{\phi} < \frac{\pi}{2}\right)$ such that

$$\det \mathbf{Q}(\hat{\phi})$$

$$= \det \left(\sum_{j,l=1}^{3} C_{ijkl}^d (m_j \cos \hat{\phi} + n_j \sin \hat{\phi})(m_l \cos \hat{\phi} + n_l \sin \hat{\phi}) \right)_{i \downarrow k \to 1,2,3} = 0.$$

Divided by $\cos^6 \hat{\phi}$, this becomes

$$\det \left(\sum_{j,l=1}^{3} C_{ijkl}^d (m_j + n_j \tan \hat{\phi})(m_l + n_l \tan \hat{\phi}) \right)_{i \downarrow k \to 1,2,3} = 0,$$

which, by (3.11), is equivalent to

$$\det[\mathbf{Q} + p(\mathbf{R} + \mathbf{R}^T) + p^2 \mathbf{T}] = 0$$

with real $p = \tan \hat{\phi}$.

When $v = v_L(\mathbf{m}, \mathbf{n})$, solution (3.4) becomes

$$\mathbf{u} = (u_1, u_2, u_3) = \mathbf{a} \, e^{-\sqrt{-1} \, k(\mathbf{m} \cdot \mathbf{x} + \tan \hat{\phi} \, \mathbf{n} \cdot \mathbf{x} - v_L t)}, \tag{3.126}$$

where $\mathbf{a} \neq \mathbf{0}$ satisfies (3.10) with $p = \tan \hat{\phi}$. This solution represents a plane wave called a limiting body wave; it has velocity

$$\frac{v_L}{|\mathbf{m} + \tan \hat{\phi} \, \mathbf{n}|} = v_L |\cos \hat{\phi}|$$

 Springer

and has direction of propagation

$$\mathbf{m} + \tan \hat{\phi}\, \mathbf{n} = \frac{\mathbf{m} \cos \hat{\phi} + \mathbf{n} \sin \hat{\phi}}{\cos \hat{\phi}}. \qquad (3.127)$$

The direction of the vector (3.127) coincides with that of $\tilde{\mathbf{m}}$ in (3.16) with $\phi = \hat{\phi}\left(-\frac{\pi}{2} < \hat{\phi} < \frac{\pi}{2}\right)$. Note that the solution (3.126), unlike the surface-wave solution (3.4), no longer decays exponentially with depth beneath the surface $\mathbf{n} \cdot \mathbf{x} = 0$. Solution (3.126) is called a limiting wave solution.

Limiting wave solutions are given a useful geometric interpretation in terms of the slowness section in the $\mathbf{m}$-$\mathbf{n}$ plane.

Let $\tilde{\mathbf{m}} = \tilde{\mathbf{m}}(\phi) = (\tilde{m}_1, \tilde{m}_2, \tilde{m}_3)$ be the unit vector in (3.16). In an elastic material, the solution to (3.3) of the form

$$\boldsymbol{u} = (u_1, u_2, u_3) = \mathbf{a}\, e^{-\sqrt{-1}\, k(\tilde{\mathbf{m}} \cdot \mathbf{x} - c(\phi)\, t)} \qquad (3.128)$$

represents a body wave which has direction of propagation $\tilde{\mathbf{m}}$, wave number k, velocity $c(\phi)$, and polarization $\mathbf{a} = \mathbf{a}(\phi)$. Substituting this into (3.3) leads to

$$\left(\sum_{j,l=1}^{3} C_{ijkl}\, \tilde{m}_j \tilde{m}_l\right)_{i\downarrow k \to 1,2,3} \mathbf{a} = \rho\, c(\phi)^2 \mathbf{a}, \qquad (3.129)$$

which implies that $\rho\, c(\phi)^2$ and $\mathbf{a}(\phi)$ are an eigenvalue and an eigenvector of the real symmetric matrix

$$\left(\sum_{j,l=1}^{3} C_{ijkl}\, \tilde{m}_j \tilde{m}_l\right)_{i\downarrow k \to 1,2,3}, \qquad (3.130)$$

respectively. We call this matrix the acoustical tensor.

As in (3.23), let

$$\lambda_i(\phi) \ (i = 1, 2, 3), \qquad 0 < \lambda_1(\phi) \le \lambda_2(\phi) \le \lambda_3(\phi)$$

be the eigenvalues of the acoustical tensor (3.130). Corresponding to these eigenvalues, there exist three body waves which have direction of propagation $\tilde{\mathbf{m}}$, velocity

$$c_i(\phi) = \sqrt{\frac{\lambda_i(\phi)}{\rho}} \ (i = 1, 2, 3), \qquad 0 < c_1(\phi) \le c_2(\phi) \le c_3(\phi) \qquad (3.131)$$

and polarizations $\mathbf{a}_i = \mathbf{a}_i(\phi) \in \mathbb{R}^3$ $(i = 1, 2, 3)$, which are mutually orthogonal.

Definition 3.24 For given orthogonal unit vectors $\mathbf{m}$ and $\mathbf{n}$ in $\mathbb{R}^3$, let $\tilde{\mathbf{m}} = \tilde{\mathbf{m}}(\phi)$ be given by (3.16) and let $c_i(\phi)$ $(i = 1, 2, 3)$ be the velocity (3.131) of the body wave which propagates in the direction of $\tilde{\mathbf{m}}$. Then the slowness section in the $\mathbf{m}$-$\mathbf{n}$ plane consists of the three closed curves which are generated by the radius vectors

$$\frac{1}{c_i(\phi)}\, \tilde{\mathbf{m}}, \qquad i = 1, 2, 3 \qquad (-\pi < \phi \le \pi).$$

Since the acoustical tensor (3.130) is π-periodic in ϕ, the slowness section is centrosymmetric with respect to the origin. The curve corresponding to the slowest velocity $c_1(\phi)$ defines the silhouette of the slowness section and is called the outer profile.

From (3.25), (3.26) and (3.131) it follows that

$$v_L = v_L(\mathbf{m}, \mathbf{n}) = \min_{-\frac{\pi}{2} < \phi < \frac{\pi}{2}} \frac{c_1(\phi)}{\cos\phi}, \tag{3.132}$$

and hence

$$v_L^{-1} = \max_{-\frac{\pi}{2} < \phi < \frac{\pi}{2}} \frac{1}{c_1(\phi)} \cos\phi. \tag{3.133}$$

Since ϕ is the angle of rotation about $\mathbf{m} \times \mathbf{n}$ between the $\mathbf{m}$-axis and the vector $\tilde{\mathbf{m}}$,

$$\frac{1}{c_1(\phi)} \cos\phi$$

is the projection on the $\mathbf{m}$-axis of the point on the outer profile. Then by (3.133), the limiting slowness $v_L(\mathbf{m}, \mathbf{n})^{-1}$ is the absolute maximum of the set of such projections. Thus, we obtain a useful construction of $v_L(\mathbf{m}, \mathbf{n})$ in terms of the slowness section.

Proposition 3.25 *In the $\mathbf{m}$-$\mathbf{n}$ plane, let L be a line parallel to the $\mathbf{n}$-axis approaching the slowness section from the right and make the first tangential contact with the outer profile at some point T. The limiting slowness $v_L(\mathbf{m}, \mathbf{n})^{-1}$ is the projection of T on the $\mathbf{m}$-axis. Let $\hat{\phi}$ be the angle between $\overrightarrow{OT}$ and the $\mathbf{m}$-axis $\left(-\frac{\pi}{2} < \hat{\phi} < \frac{\pi}{2}\right)$. Then*

$$v_L(\mathbf{m}, \mathbf{n})^{-1} = \max_{-\frac{\pi}{2} < \phi < \frac{\pi}{2}} \frac{1}{c_1(\phi)} \cos\phi = \frac{1}{c_1(\hat{\phi})} \cos\hat{\phi} \tag{3.134}$$

(see Fig. 1). The corresponding body wave represented by (3.128) propagates in the direction of the radius vector $\overrightarrow{OT}$ with the velocity $c_1(\hat{\phi})$ and its polarization is an eigenvector of the acoustical tensor (3.130) at $\phi = \hat{\phi}$ pertaining to the smallest eigenvalue $\rho\, c_1(\hat{\phi})^2$.

Here we show that the limiting wave solution (3.126), which is derived from the transonic limit $v \uparrow v_L$ of the surface-wave solution (3.4), is essentially the same as the body-wave solution (3.128) with $c(\phi) = c_1(\phi)$ and $\phi = \hat{\phi}$.

Noting (3.127), (3.134) and $-\frac{\pi}{2} < \hat{\phi} < \frac{\pi}{2}$, we may write (3.126) as

$$\mathbf{u} = \mathbf{a} \exp\left\{-\sqrt{-1}\,k\left[\frac{\mathbf{m}\cos\hat{\phi} + \mathbf{n}\sin\hat{\phi}}{\cos\hat{\phi}} \cdot \mathbf{x} - \frac{c_1(\hat{\phi})}{\cos\hat{\phi}} t\right]\right\}$$

$$= \mathbf{a} \exp\left\{-\sqrt{-1}\frac{1}{\cos\hat{\phi}}\,k\left(\tilde{\mathbf{m}}(\hat{\phi}) \cdot \mathbf{x} - c_1(\hat{\phi})t\right)\right\},$$

the exponent of which coincides with that of (3.128) with $c(\phi) = c_1(\phi)$ and $\phi = \hat{\phi}$ except for a constant multiplier $\frac{1}{\cos\hat{\phi}}$ in the wave number, and we see that the equation for $\mathbf{a}$ in the above, i.e., (3.10) with $p = \tan\hat{\phi}$, namely,

$$\left(\sum_{j,l=1}^{3} C_{ijkl}^{d}\,(m_j + n_j \tan\hat{\phi})(m_l + n_l \tan\hat{\phi})\right)_{i\downarrow k \to 1,2,3} \mathbf{a} = \mathbf{0},$$

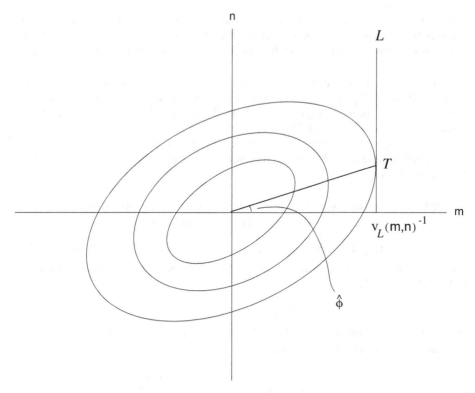

Fig. 1 Slowness section and limiting velocity

becomes, by (3.6), (3.16) and (3.134),

$$
\left[\left(\sum_{j,l=1}^{3} C_{ijkl}\,(m_j + n_j \tan\hat{\phi})(m_l + n_l \tan\hat{\phi})\right)_{i\downarrow k\to 1,2,3} - \rho\,v_L^2\,\mathbf{I}\right]\mathbf{a}
$$

$$
= \left[\left(\sum_{j,l=1}^{3} C_{ijkl}\,(m_j + n_j \tan\hat{\phi})(m_l + n_l \tan\hat{\phi})\right)_{i\downarrow k\to 1,2,3} - \rho\,\frac{c_1(\hat{\phi})^2}{\cos^2\hat{\phi}}\,\mathbf{I}\right]\mathbf{a}
$$

$$
= \frac{1}{\cos^2\hat{\phi}}\left[\left(\sum_{j,l=1}^{3} C_{ijkl}\,\tilde{m}_j(\hat{\phi})\tilde{m}_l(\hat{\phi})\right)_{i\downarrow k\to 1,2,3} - \rho\,c_1(\hat{\phi})^2\,\mathbf{I}\right]\mathbf{a} = \mathbf{0}. \qquad (3.135)
$$

The last equation is the same as (3.129) with $c(\phi) = c_1(\phi)$ and $\phi = \hat{\phi}$. Hence, we also call solution (3.128) with $c(\phi) = c_1(\phi)$ and $\phi = \hat{\phi}$ a limiting wave solution.

Remark 3.26 Given orthogonal unit vectors $\mathbf{m}$ and $\mathbf{n}$, the limiting body wave is not necessarily unique. In fact, two or three $\hat{\phi}_i$ may satisfy (3.134). In this case, different points of tangency with L, i.e., different T_i, are generated on the outer profile. Then there arise limiting body waves which have different directions of propagation $\overrightarrow{OT_i}$.

Also it may happen that $c_1(\hat{\phi}) = c_2(\hat{\phi})$ or $c_1(\hat{\phi}) = c_2(\hat{\phi}) = c_3(\hat{\phi})$. In this case, two or three closed curves in the slowness section make tangential contact at one T. Then there appear limiting body waves which have different (orthogonal) polarizations.

In the **m-n** plane, the transonic state is the state of the slowness section at the point of the first tangential contact with the vertical line L which approaches the slowness section from the right in the plane. To discuss the existence of Rayleigh waves, Chadwick and Smith [22] classified the transonic state into six possible types according to the number of points of tangency T and the number of closed curves participating in each tangential contact. We shall return to this point later in Subsection 3.6.4.

Example The slowness section of an isotropic material is constituted by three circles centered at the origin, two of which are of the same radius $\sqrt{\frac{\rho}{\mu}}$ and the other of radius $\sqrt{\frac{\rho}{\lambda+2\mu}}$. These are independent of the choice of **m** and **n**. This fact follows from the consideration below. The acoustical tensor (3.130) of isotropic elasticity is given by (3.68). Then by (3.72), it becomes

$$(\lambda + 2\mu)\tilde{\mathbf{m}} \otimes \tilde{\mathbf{m}} + \mu(\tilde{\mathbf{n}} \otimes \tilde{\mathbf{n}} + \boldsymbol{\ell} \otimes \boldsymbol{\ell}),$$

the eigenvalues of which are $\lambda + 2\mu$ (simple) and μ (double). It follows from (3.131) that the velocity of the body waves are

$$c_1(\phi) = c_2(\phi) = \sqrt{\frac{\mu}{\rho}}, \qquad c_3(\phi) = \sqrt{\frac{\lambda + 2\mu}{\rho}}$$

for all ϕ.[43]

Now let us compute the limiting body waves in an isotropic material. From (3.134) or Fig. 1 it follows that $\hat{\phi} = 0$ and $v_L^{\text{Iso}} = \sqrt{\frac{\mu}{\rho}}$. Therefore, two limiting body waves propagate in the direction of **m** with the velocity $\sqrt{\frac{\mu}{\rho}}$, and their polarizations $\mathbf{a}_\alpha$ ($\alpha = 1, 2$) satisfy (3.129);

$$\left[(\lambda + 2\mu)\mathbf{m} \otimes \mathbf{m} + \mu(\mathbf{n} \otimes \mathbf{n} + \boldsymbol{\ell} \otimes \boldsymbol{\ell})\right] \mathbf{a}_\alpha = \mu \, \mathbf{a}_\alpha,$$

which from (3.72) becomes

$$\left[(\lambda + \mu)\mathbf{m} \otimes \mathbf{m}\right]\mathbf{a}_\alpha = \mathbf{0}.$$

Hence by (3.65) we may take

$$\mathbf{a}_1 = \mathbf{m} \times \mathbf{n}, \qquad \mathbf{a}_2 = \mathbf{n}.$$

These are identical to those in (3.36) when $p_1 = \tan\hat{\phi} = 0$.

---

[43] By (3.65), the corresponding polarizations are taken to be

$$\mathbf{a}_1 = \boldsymbol{\ell} \, (= \tilde{\mathbf{m}} \times \tilde{\mathbf{n}} = \mathbf{m} \times \mathbf{n}), \qquad \mathbf{a}_2 = \tilde{\mathbf{n}}, \qquad \mathbf{a}_3 = \tilde{\mathbf{m}}.$$

The first two polarizations pertain to shear waves and the last one pertains to a longitudinal wave (see (3.128)).

 Springer

Finally, for later use, we compute the traction on the surface $\mathbf{n} \cdot \mathbf{x} = 0$ produced by the limiting wave solution.

Lemma 3.27 *The traction on the surface* $\mathbf{n} \cdot \mathbf{x} = 0$ *produced by the limiting wave solution* (3.128) *with* $c(\phi) = c_1(\phi)$ *and* $\phi = \hat{\phi}$ *is given by*

$$
t_L = -\sqrt{-1}\, k \cos \hat{\phi} \left[\mathbf{R}(0)^T + p\, \mathbf{T}(0) \right]_{p=\tan \hat{\phi}} \mathbf{a}(\hat{\phi})\, e^{-\sqrt{-1}\, k \left(\widetilde{\mathbf{m}}(\hat{\phi}) \cdot \mathbf{x} - c_1(\hat{\phi}) t \right)} \quad (3.136)
$$

$$
= -\sqrt{-1}\, k \cos \hat{\phi}\, \mathbf{R}(\hat{\phi})^T \Big|_{v=v_L} \mathbf{a}(\hat{\phi})\, e^{-\sqrt{-1}\, k \left(\widetilde{\mathbf{m}}(\hat{\phi}) \cdot \mathbf{x} - c_1(\hat{\phi}) t \right)}, \quad (3.137)
$$

where $\hat{\phi}$ *is the angle in* (3.134), $\mathbf{R}(\phi)$ *and* $\mathbf{T}(\phi)$ *are the* 3×3 *matrices in* (3.17), $\mathbf{a}(\hat{\phi})$ *is an eigenvector of the acoustical tensor* (3.130) *at* $\phi = \hat{\phi}$ *associated with its smallest eigenvalue* $\rho\, c_1(\hat{\phi})^2$ *and* $\widetilde{\mathbf{m}}(\phi)$ *is the unit vector in* (3.16).

Proof For simplicity we suppress the dependence of the vectors $\widetilde{\mathbf{m}}(\hat{\phi})$ and $\widetilde{\mathbf{n}}(\hat{\phi})$ on $\hat{\phi}$.

By definition (1.24), the traction on the surface $\mathbf{n} \cdot \mathbf{x} = 0$ produced by (3.128) with $c(\phi) = c_1(\phi)$ and $\phi = \hat{\phi}$ is

$$
t_L = -\sqrt{-1}\, k \left(\sum_{j,l=1}^{3} C_{ijkl}\, n_j \widetilde{m}_l \right)_{i \downarrow k \to 1,2,3} \mathbf{a}(\hat{\phi})\, e^{-\sqrt{-1}\, k \left(\widetilde{\mathbf{m}} \cdot \mathbf{x} - c_1(\hat{\phi}) t \right)}. \quad (3.138)
$$

The first equality (3.136) is obvious, since (3.9) implies that

$$
\left[\mathbf{R}(0)^T + p\, \mathbf{T}(0) \right]_{p=\tan \hat{\phi}} = \left(\sum_{j,l=1}^{3} C_{ijkl}\, (n_j m_l + n_j n_l \tan \hat{\phi}) \right)_{i \downarrow k \to 1,2,3},
$$

and then it follows from (3.16) that

$$
\cos \hat{\phi} \left[\mathbf{R}(0)^T + p\, \mathbf{T}(0) \right]_{p=\tan \hat{\phi}}
$$

$$
= \left(\sum_{j,l=1}^{3} C_{ijkl}\, n_j (m_l \cos \hat{\phi} + n_l \sin \hat{\phi}) \right)_{i \downarrow k \to 1,2,3} = \left(\sum_{j,l=1}^{3} C_{ijkl}\, n_j \widetilde{m}_l \right)_{i \downarrow k \to 1,2,3},
$$

which is equal to the matrix in the right hand side of (3.138).

Now we prove the second equality (3.137). From (3.58) we get

$$
\cos \hat{\phi}\, \mathbf{R}(\hat{\phi})^T \Big|_{v=v_L} \mathbf{a}(\hat{\phi})
$$

$$
= \left[\cos \hat{\phi} \left(\sum_{j,l=1}^{3} C_{ijkl}\, \widetilde{n}_j \widetilde{m}_l \right)_{i \downarrow k \to 1,2,3} + \rho\, v_L^2 \cos^2 \hat{\phi} \sin \hat{\phi}\, \mathbf{I} \right] \mathbf{a}(\hat{\phi}). \quad (3.139)
$$

Since (3.135) and (3.16) imply that

$$
\rho\, v_L^2\, \mathbf{a}(\hat{\phi}) = \left(\sum_{j,l=1}^{3} C_{ijkl}\, (m_j + n_j \tan \hat{\phi})(m_l + n_l \tan \hat{\phi}) \right)_{i \downarrow k \to 1,2,3} \mathbf{a}(\hat{\phi})
$$

Springer

and

$$\rho \, v_L^2 \cos^2 \hat{\phi} \, \sin \hat{\phi} \, \mathbf{a}(\hat{\phi}) = \sin \hat{\phi} \left(\sum_{j,l=1}^{3} C_{ijkl} \, \tilde{m}_j \tilde{m}_l \right)_{i \downarrow k \to 1,2,3} \mathbf{a}(\hat{\phi}),$$

(3.139) becomes

$$\cos \hat{\phi} \, \mathbf{R}(\hat{\phi})^T \Big|_{v=v_L} \mathbf{a}(\hat{\phi}) = \left(\sum_{j,l=1}^{3} C_{ijkl} \left(\tilde{n}_j \cos \hat{\phi} + \tilde{m}_j \sin \hat{\phi} \right) \tilde{m}_l \right)_{i \downarrow k \to 1,2,3} \mathbf{a}(\hat{\phi}).$$

Again (3.16) implies that

$$\tilde{n}_j \cos \hat{\phi} + \tilde{m}_j \sin \hat{\phi} = n_j \left(\cos^2 \hat{\phi} + \sin^2 \hat{\phi} \right) = n_j.$$

This proves (3.137).   □

In the preceding sections we have seen that the matrix $\mathbf{S}_3 = \mathbf{S}_3(v)$ is useful for the study of Rayleigh waves. In the next subsection, as a continuation of this approach, we shall give a criterion for the existence of Rayleigh waves which is based on the behavior of $\mathbf{S}_3(v)$ in the transonic limit $v \uparrow v_L$. In Subsection 3.6.3 we give a criterion for the existence of Rayleigh waves which is based on the behavior of the surface impedance $\mathbf{Z}(v)$ in the transonic limit $v \uparrow v_L$. Finally, in Subsection 3.6.4, we give an existence criterion for Rayleigh waves based on Chadwick and Smith's classification [22] of the transonic state.

3.6.2 Existence Criterion Based on $\mathbf{S}_3$

It is known that the real symmetric matrix $\mathbf{S}_3(0)$ is negative definite. For the proof, which is long, we refer to Section 5.C of [22], [31, 76] and Section 7.6 of [77].[44] Since the negative definiteness of $\mathbf{S}_3(0)$ implies that all its eigenvalues are negative, it follows from (2) of Lemma 3.15 that all the eigenvalues of $\mathbf{S}_3(v)$ increase monotonically with v in the interval $0 \leq v < v_L$ from their negative values at $v = 0$.

Now we give a criterion for the existence of Rayleigh waves based on the behavior of $\mathbf{S}_3(v)$ in the transonic limit $v \uparrow v_L$.

Theorem 3.28 *For given orthogonal unit vectors* $\mathbf{m}$ *and* $\mathbf{n}$ *in* $\mathbb{R}^3$, *let* $\mathbf{S}_3 = \mathbf{S}_3(v)$ *be the* 3×3 *real symmetric matrix defined by* (3.39) *for* $0 \leq v < v_L$. *Let* $v_L = v_L(\mathbf{m}, \mathbf{n})$ *be the limiting velocity given by* (3.20) *(or equivalently by* (3.132)), *and let*

$$\mathbf{S}_3(v_L) = \lim_{v \uparrow v_L} \mathbf{S}_3(v).$$

A necessary and sufficient condition for the existence of the Rayleigh waves in the half-space $\mathbf{n} \cdot \mathbf{x} \leq 0$ *which propagate along the surface* $\mathbf{n} \cdot \mathbf{x} = 0$ *in the direction of* $\mathbf{m}$ *with*

---

[44]The positive definiteness of $\mathbf{S}_2(v)$ $(0 \leq v < v_L)$ and that of $\frac{d}{dv}\mathbf{S}_3$ $(0 < v < v_L)$ are proved under the strong ellipticity condition (1.79) because only the positive definiteness of $\mathbf{T}(\phi)$ for all ϕ is needed there (see Lemma 3.5 and Lemma 3.15). On the other hand, the more stringent strong convexity condition (1.7) is needed for the negative definiteness of $\mathbf{S}_3(0)$ (cf. the references cited).

the phase velocity in the subsonic range $0 < v < v_L$ is

$$\operatorname{tr} \mathbf{S}_3(v_L) > 0, \,^{45} \tag{3.140}$$

where tr *denotes the trace of a square matrix.*[46]

Example In isotropic elasticity, we see from (3.21) and (3.79) that

$$\operatorname{tr} \mathbf{S}_3 \left(v_L^{\mathrm{Iso}} \right) = +\infty.^{47}$$

Hence a Rayleigh wave exists for any **m** and **n**.

Proof of sufficiency Suppose that (3.140) holds. Since the trace of a square matrix is equal to the sum of its eigenvalues, at least one of the eigenvalues of $\mathbf{S}_3(v_L)$ is positive. Since all the eigenvalues of $\mathbf{S}_3(0)$ are negative, by the continuous dependence of the eigenvalues of $\mathbf{S}_3(v)$ on v, at least one of them vanishes at some v $(0 < v < v_L)$. Lemma 3.11 and Theorem 3.14 imply that this v is v_R.

Before proving necessity, we investigate the behavior of $\mathbf{S}_i(v)$ $(i = 1, 2, 3)$ in the transonic limit $v \uparrow v_L$.

Lemma 3.29 *Let $\mathbf{R}(\phi)$ and $\mathbf{T}(\phi)$ be the 3×3 matrices in (3.17), and let $\mathbf{a}_i(\phi)$ $(i = 1, 2, 3)$ be orthonormal eigenvectors of the acoustical tensor (3.130) corresponding to the eigenvalues $\lambda_i(\phi) = \rho \, c_i(\phi)^2$ $(i = 1, 2, 3, \; 0 < \lambda_1(\phi) \le \lambda_2(\phi) \le \lambda_3(\phi))$, which form a real orthonormal basis in $\mathbb{R}^3$. Then for $0 \le v < v_L$,*

$$\mathbf{S}_1(v) = \frac{1}{\pi} \sum_{i=1}^{3} \frac{1}{\rho} \int_{-\frac{\pi}{2}}^{\frac{\pi}{2}} \frac{\mathbf{a}_i(\phi) \otimes \mathbf{R}(\phi)^T \mathbf{a}_i(\phi)}{c_i(\phi)^2 - v^2 \cos^2 \phi} \, d\phi, \tag{3.141}$$

$$\mathbf{S}_2(v) = \frac{1}{\pi} \sum_{i=1}^{3} \frac{1}{\rho} \int_{-\frac{\pi}{2}}^{\frac{\pi}{2}} \frac{\mathbf{a}_i(\phi) \otimes \mathbf{a}_i(\phi)}{c_i(\phi)^2 - v^2 \cos^2 \phi} \, d\phi, \tag{3.142}$$

$$\mathbf{S}_3(v) = \frac{1}{\pi} \int_{-\frac{\pi}{2}}^{\frac{\pi}{2}} -\mathbf{T}(\phi) \, d\phi + \frac{1}{\pi} \sum_{i=1}^{3} \frac{1}{\rho} \int_{-\frac{\pi}{2}}^{\frac{\pi}{2}} \frac{\mathbf{R}(\phi)^T \mathbf{a}_i(\phi) \otimes \mathbf{R}(\phi)^T \mathbf{a}_i(\phi)}{c_i(\phi)^2 - v^2 \cos^2 \phi} \, d\phi, \tag{3.143}$$

where the symbol $\otimes$ denotes the tensor product of two vectors (see the footnote of (3.64)).

---

[45] Here the case $\operatorname{tr} \mathbf{S}_3(v_L) = +\infty$ is included.

[46] The part on necessity in the theorem can be proved under the condition that the outer profile of the slowness section has non-zero curvature at T, i.e., at the point of tangency with L. In this section, however, we do not consider that special case where the outer profile has zero curvature at T. For the work on this special case, we refer to [13] (see also the comment after Remark 3.39). Hence in the proof of necessity below we assume that the outer profile of the slowness section has non-zero curvature at T.

[47] Note that the coefficient of $\mathbf{m} \otimes \mathbf{m}$ in (3.79) blows up when $V \uparrow \mu$.

Proof From (3.19), (3.39) and π-periodicity in ϕ of the integrands of $\mathbf{S}_i(v)$ ($i = 1, 2, 3$) it follows that

$$\mathbf{S}_1(v) = \frac{1}{\pi} \int_{-\frac{\pi}{2}}^{\frac{\pi}{2}} \mathbf{Q}(\phi)^{-1} \mathbf{R}(\phi) \, d\phi, \qquad \mathbf{S}_2(v) = \frac{1}{\pi} \int_{-\frac{\pi}{2}}^{\frac{\pi}{2}} \mathbf{Q}(\phi)^{-1} \, d\phi,$$

$$\mathbf{S}_3(v) = \frac{1}{\pi} \int_{-\frac{\pi}{2}}^{\frac{\pi}{2}} -\mathbf{T}(\phi) + \mathbf{R}(\phi)^T \mathbf{Q}(\phi)^{-1} \mathbf{R}(\phi) \, d\phi.$$

Recalling that the eigenvalues of $\mathbf{Q}(\phi)$ are given by (3.24), we see that the eigenvalues of $\mathbf{Q}(\phi)^{-1}$ are

$$\frac{1}{\lambda_i(\phi) - \rho \, v^2 \cos^2 \phi} = \frac{1}{\rho(c_i(\phi)^2 - v^2 \cos^2 \phi)}, \qquad i = 1, 2, 3,$$

whereas by (3.22), eigenvectors of $\mathbf{Q}(\phi)$, and hence those of $\mathbf{Q}(\phi)^{-1}$, remain $\mathbf{a}_i(\phi)$ ($i = 1, 2, 3$). Hence the spectral representation[48] for $\mathbf{Q}(\phi)^{-1}$ is

$$\mathbf{Q}(\phi)^{-1} = \sum_{i=1}^{3} \frac{\mathbf{a}_i(\phi) \otimes \mathbf{a}_i(\phi)}{\rho(c_i(\phi)^2 - v^2 \cos^2 \phi)}.$$

Then from the formulas in tensor algebra[49] we obtain the lemma.   □

Depending on the transonic state, there are different scenarios regarding limiting body waves (cf. Remark 3.26 and the subsequent paragraph; see also the footnote of Fig. 3). Here it is convenient to classify the transonic state according to the property of the tractions produced by limiting body waves.

Definition 3.30 Let $\mathbf{m}$ and $\mathbf{n}$ be orthogonal unit vectors in $\mathbb{R}^3$. The transonic state is *normal* if at least one limiting wave solution produces non-zero tractions on the surface $\mathbf{n} \cdot \mathbf{x} = 0$. The transonic state is *exceptional* if every limiting wave solution produces no tractions on the surface $\mathbf{n} \cdot \mathbf{x} = 0$.

Now we sum up the behavior of $\mathbf{S}_i(v)$ ($i = 1, 2, 3$) in the transonic limit $v \uparrow v_L$.

Lemma 3.31 (1) *When the transonic state is normal, at least one eigenvalue of $\mathbf{S}_3(v)$ approaches positive infinity as $v \uparrow v_L$. When the transonic state is exceptional, all the eigenvalues of $\mathbf{S}_3(v)$ remain bounded as $v \uparrow v_L$.*
(2) *When the transonic state is normal, at least one component of $\mathbf{S}_1(v)$ becomes unbounded as $v \uparrow v_L$. When the transonic state is exceptional, all the components of $\mathbf{S}_1(v)$ remain bounded as $v \uparrow v_L$.*

---

[48]Let $\mathbf{e}_i$ ($i = 1, 2, 3$) be a set of orthonormal eigenvectors in $\mathbb{R}^3$ (resp. $\mathbb{C}^3$) of a real symmetric (resp. an Hermitian) 3×3 matrix $\mathbf{M}$ associated with the eigenvalues λ_i ($i = 1, 2, 3$). Then

$$\mathbf{M} = \sum_{i=1}^{3} \lambda_i \, \mathbf{e}_i \otimes \mathbf{e}_i. \tag{3.144}$$

[49]For two $\mathbb{R}^3$-vectors $\mathbf{a}, \mathbf{b}$ and a 3×3 real matrix $\mathbf{M}$,

$$(\mathbf{a} \otimes \mathbf{b})\mathbf{M} = \mathbf{a} \otimes \mathbf{M}^T \mathbf{b}, \qquad \mathbf{M}(\mathbf{a} \otimes \mathbf{b}) = \mathbf{M}\mathbf{a} \otimes \mathbf{b}.$$

(3)   *At least one eigenvalue of $S_2(v)$ approaches positive infinity as $v \uparrow v_L$ regardless of whether the transonic state is normal or exceptional.*

Proof (1)   We show that in the case where the transonic state is normal, there exists a non-zero vector $\mathbf{b} \in \mathbb{R}^3$ such that $\mathbf{b} \cdot S_3(v)\mathbf{b} \longrightarrow +\infty$ as $v \uparrow v_L$, and that in the case where the transonic state is exceptional, for any non-zero vector $\mathbf{b} \in \mathbb{R}^3$, $\mathbf{b} \cdot S_3(v)\mathbf{b}$ remaims bounded as $v \uparrow v_L$. Note that from (3.143) and (3.65) it follows that

$$\mathbf{b} \cdot S_3(v)\mathbf{b} = \frac{1}{\pi} \int_{-\frac{\pi}{2}}^{\frac{\pi}{2}} -\mathbf{b} \cdot \mathbf{T}(\phi)\mathbf{b}\, d\phi + \frac{1}{\pi} \sum_{i=1}^{3} \frac{1}{\rho} \int_{-\frac{\pi}{2}}^{\frac{\pi}{2}} \frac{\left(\mathbf{b} \cdot \mathbf{R}(\phi)^T \mathbf{a}_i(\phi)\right)^2}{c_i(\phi)^2 - v^2 \cos^2 \phi}\, d\phi.$$

The first integral on the right hand side is bounded as $v \uparrow v_L$. Hence we investigate the remaining integrals.

The proof is long, so we first give an outline. It follows from (3.134) that

$$c_1(\phi) - v_L \cos\phi \quad \downarrow \quad 0 \quad \text{as} \quad \phi \longrightarrow \hat{\phi}.$$

Hence, for the index i such that $c_i(\hat{\phi}) > c_1(\hat{\phi})$, the integral

$$\int_{-\frac{\pi}{2}}^{\frac{\pi}{2}} \frac{\left(\mathbf{b} \cdot \mathbf{R}(\phi)^T \mathbf{a}_i(\phi)\right)^2}{c_i(\phi)^2 - v^2 \cos^2 \phi}\, d\phi \tag{3.145}$$

remains bounded as $v \uparrow v_L$. Moreover, for $i = 1$, and for the index i such that $c_i(\hat{\phi}) = c_1(\hat{\phi})$, if the integral (3.145) diverges as $v \uparrow v_L$, then this must arise only from the contribution of the neighborhood of the angle $\hat{\phi}$.[50]

In the case where the transonic state is normal, from (3.137) there exist $\mathbf{a}_i(\phi)$ and $c_i(\phi)$ of the body wave solution (3.128) such that $c_i(\hat{\phi}) = c_1(\hat{\phi})$ and

$$\mathbf{R}(\hat{\phi})^T \Big|_{v=v_L} \mathbf{a}_i(\hat{\phi}) \neq \mathbf{0}. \tag{3.146}$$

Then, for $\mathbf{b} \in \mathbb{R}^3$ such that

$$\left(\mathbf{b} \cdot \mathbf{R}(\hat{\phi})^T \Big|_{v=v_L} \mathbf{a}_i(\hat{\phi})\right) \neq \mathbf{0}, \tag{3.147}$$

we shall show that

$$\int_{\hat{\phi}-\delta}^{\hat{\phi}+\delta} \frac{\left(\mathbf{b} \cdot \mathbf{R}(\phi)^T \mathbf{a}_i(\phi)\right)^2}{c_i(\phi)^2 - v^2 \cos^2 \phi}\, d\phi \longrightarrow +\infty \quad \text{as} \quad v \uparrow v_L$$

for some small $\delta > 0$.

In the case where the transonic state is exceptional, let $\mathbf{a}_i(\phi)$ and $c_i(\phi)$ be those of the body wave solution (3.128) such that $c_i(\hat{\phi}) = c_1(\hat{\phi})$. Then for such i, (3.137) implies that

$$\mathbf{R}(\hat{\phi})^T \Big|_{v=v_L} \mathbf{a}_i(\hat{\phi}) = \mathbf{0}. \tag{3.148}$$

---

[50] By Remark 3.26, two or three $\hat{\phi}$ may occur. In this case we consider the contribution of each neighborhood of $\hat{\phi}$.

Then for any $\mathbf{b} \in \mathbb{R}^3$,

$$\left(\mathbf{b} \cdot \mathbf{R}(\hat{\phi})^T \Big|_{v=v_L} \mathbf{a}_i(\hat{\phi}) \right) = 0. \tag{3.149}$$

In this case, we shall show that

$$\int_{\hat{\phi}-\delta}^{\hat{\phi}+\delta} \frac{\left(\mathbf{b} \cdot \mathbf{R}(\phi)^T \mathbf{a}_i(\phi) \right)^2}{c_i(\phi)^2 - v^2 \cos^2 \phi} \, d\phi \; < \infty \qquad \text{as} \quad v \uparrow v_L$$

for some small $\delta > 0$.

Now we proceed to the details. From (3.132) we put

$$\frac{d}{d\phi} \left(\frac{c_i(\phi)}{\cos \phi} \right) = 0, \qquad \frac{d^2}{d\phi^2} \left(\frac{c_i(\phi)}{\cos \phi} \right) = G > 0^{51} \qquad \text{at } \phi = \hat{\phi}. \tag{3.150}$$

Since $\frac{d^2}{d\phi^2} \left(\frac{c_i(\phi)}{\cos \phi} \right)$ in a neighborhood of $\phi = \hat{\phi}$ is bounded from above by a positive constant and from below by another positive constant, Taylor's theorem gives

$$v_L + C(\phi - \hat{\phi})^2 \leq \frac{c_i(\phi)}{\cos \phi} \leq v_L + C'(\phi - \hat{\phi})^2 \qquad (\, |\phi - \hat{\phi}| \leq \delta \,)$$

for some small $\delta > 0$. Here and hereafter we use C's (or C''s) to denote positive constants depending on v_L, δ (and δ') without distinguishing between them. Then,

$$v_L^2 + C(\phi - \hat{\phi})^2 \leq \frac{c_i(\phi)^2}{\cos^2 \phi} \leq v_L^2 + C'(\phi - \hat{\phi})^2 \qquad (\, |\phi - \hat{\phi}| \leq \delta \,)$$

and

$$C \, (\phi - \hat{\phi})^2 \cos^2 \phi \leq c_i(\phi)^2 - v_L^2 \cos^2 \phi \leq C' \, (\phi - \hat{\phi})^2 \cos^2 \phi \qquad (\, |\phi - \hat{\phi}| \leq \delta \,). \tag{3.151}$$

In the case where the transonic state is normal, (3.147) implies that for small $\delta, \delta' > 0$,

$$\left| \left(\mathbf{b} \cdot \mathbf{R}(\phi)^T \mathbf{a}_i(\phi) \right) \right| > C > 0 \qquad (\, |\phi - \hat{\phi}| \leq \delta, \quad v_L - \delta' \leq v \leq v_L \,).$$

Hence we get

$$\int_{\hat{\phi}-\delta}^{\hat{\phi}+\delta} \frac{\left(\mathbf{b} \cdot \mathbf{R}(\phi)^T \mathbf{a}_i(\phi) \right)^2}{c_i(\phi)^2 - v^2 \cos^2 \phi} \, d\phi \geq C \int_{\hat{\phi}-\delta}^{\hat{\phi}+\delta} \frac{1}{c_i(\phi)^2 - v_L^2 \cos^2 \phi + (v_L^2 - v^2) \cos^2 \phi} \, d\phi.$$

---

[51] We have assumed that the outer profile of the slowness section has non-zero curvature at T when considering necessity of the condition (3.140). Hence there exists such a G (cf. [13]).

Then by the second inequality of (3.151), the last integral can be estimated from below as

$$\int_{\hat{\phi}-\delta}^{\hat{\phi}+\delta} \frac{1}{c_i(\phi)^2 - v_L^2 \cos^2 \phi + \left(v_L^2 - v^2\right) \cos^2 \phi} \, d\phi$$

$$\geq C \int_{\hat{\phi}-\delta}^{\hat{\phi}+\delta} \frac{1}{(\phi - \hat{\phi})^2 \cos^2 \phi + \left(v_L^2 - v^2\right) \cos^2 \phi} \, d\phi$$

$$\geq C \int_{\hat{\phi}-\delta}^{\hat{\phi}+\delta} \frac{1}{(\phi - \hat{\phi})^2 + v_L^2 - v^2} \, d\phi = \frac{2C}{\sqrt{v_L^2 - v^2}} \tan^{-1} \frac{\delta}{\sqrt{v_L^2 - v^2}}.$$

The last term approaches positive infinity as $v \uparrow v_L$.

In the case where the transonic state is exceptional, the mean value theorem, combined with (3.149) and the boundedness of $\frac{d}{d\phi} \left(\mathbf{b} \cdot \mathbf{R}(\phi)^T \big|_{v=v_L} \mathbf{a}_i(\phi)\right)$ in a neighborhood of $\phi = \hat{\phi}$, implies that

$$\left| \left(\mathbf{b} \cdot \mathbf{R}(\phi)^T \big|_{v=v_L} \mathbf{a}_i(\phi)\right) \right| \leq C|\phi - \hat{\phi}| \qquad (|\phi - \hat{\phi}| \leq \delta) \qquad (3.152)$$

for small $\delta > 0$. Since it follows from (3.58) that

$$\left(\mathbf{b} \cdot \mathbf{R}(\phi)^T \mathbf{a}_i(\phi)\right) = \left(\mathbf{b} \cdot \mathbf{R}(\phi)^T \big|_{v=v_L} \mathbf{a}_i(\phi)\right) + \rho \left(v^2 - v_L^2\right) \cos \phi \sin \phi \left(\mathbf{b} \cdot \mathbf{a}_i(\phi)\right),$$

(3.152) gives

$$\left(\mathbf{b} \cdot \mathbf{R}(\phi)^T \mathbf{a}_i(\phi)\right)^2 \leq C(\phi - \hat{\phi})^2 + C' \left(v_L^2 - v^2\right)$$

$$(|\phi - \hat{\phi}| \leq \delta, \quad v_L - \delta' \leq v \leq v_L)$$

for small $\delta, \delta' > 0$. Therefore, using the first inequality of (3.151), we obtain

$$0 \leq \int_{\hat{\phi}-\delta}^{\hat{\phi}+\delta} \frac{\left(\mathbf{b} \cdot \mathbf{R}(\phi)^T \mathbf{a}_i(\phi)\right)^2}{c_i(\phi)^2 - v^2 \cos^2 \phi} \, d\phi$$

$$\leq C \int_{\hat{\phi}-\delta}^{\hat{\phi}+\delta} \frac{(\phi - \hat{\phi})^2 + v_L^2 - v^2}{c_i(\phi)^2 - v_L^2 \cos^2 \phi + \left(v_L^2 - v^2\right) \cos^2 \phi} \, d\phi$$

$$\leq C \int_{\hat{\phi}-\delta}^{\hat{\phi}+\delta} \frac{(\phi - \hat{\phi})^2 + v_L^2 - v^2}{(\phi - \hat{\phi})^2 \cos^2 \phi + \left(v_L^2 - v^2\right) \cos^2 \phi} \, d\phi, \qquad (3.153)$$

which is bounded as $v \uparrow v_L$, since $\cos \phi$ is strictly positive for $\frac{-\pi}{2} < \hat{\phi} - \delta \leq \phi \leq \hat{\phi} + \delta < \frac{\pi}{2}$.

(2) In the case where the transonic state is normal, we have (3.146). Then there exists at least one component of the matrix $\mathbf{a}_i(\phi) \otimes \mathbf{R}(\phi)^T \mathbf{a}_i(\phi)$ which does not vanish at $\phi = \hat{\phi}$ and $v = v_L$. By the same argument as in the proof of (1) for the normal transonic state, at least one component of the integrals in (3.141) diverges as $v \uparrow v_L$.

In the case where the transonic state is exceptional, we have (3.148). Then each component of the matrix $\mathbf{a}_i(\phi) \otimes \mathbf{R}(\phi)^T \mathbf{a}_i(\phi)$ vanishes at $\phi = \hat{\phi}$ and $v = v_L$. In this case we can use almost parallel arguments to the proof of (1) for

⌂ Springer

the exceptional transonic state. However, in the estimation of the integral $\int_{\hat\phi-\delta}^{\hat\phi+\delta} \frac{\mathbf{a}_i(\phi)\otimes\mathbf{R}(\phi)^T|_{v=v_L}\mathbf{a}_i(\phi)}{c_i(\phi)^2-v^2\cos^2\phi}\,d\phi$, unlike (3.153) the numerator has a first-order term in $\phi-\hat\phi$ and the integral can not be estimated from above easily as in (3.153). To solve this, by Taylor's theorem, we expand $\frac{\mathbf{a}_i(\phi)\otimes\mathbf{R}(\phi)^T|_{v=v_L}\mathbf{a}_i(\phi)}{\cos^2\phi}$ around $\phi=\hat\phi$ up to the second-order term in $\phi-\hat\phi$ and $\frac{c_i(\phi)}{\cos\phi}$ up to the third-order term in $\phi-\hat\phi$, and appeal to the equality

$$\int_{\hat\phi-\delta}^{\hat\phi+\delta} \frac{\phi-\hat\phi}{2v_L G(\phi-\hat\phi)^2 + v_L^2 - v^2}\,d\phi = 0.$$

Then we show that the remaining terms are bounded as $v \uparrow v_L$. The details are left as an exercise.

(3) For non-zero vector $\mathbf{b} \in \mathbb{R}^3$,

$$\mathbf{b}\cdot\mathbf{S}_2(v)\mathbf{b} = \frac{1}{\pi}\sum_{i=1}^{3}\frac{1}{\rho}\int_{-\frac{\pi}{2}}^{\frac{\pi}{2}} \frac{(\mathbf{b}\cdot\mathbf{a}_i(\phi))^2}{c_i(\phi)^2 - v^2\cos^2\phi}\,d\phi.$$

Then, for $\mathbf{b}$ such that $\mathbf{b}\cdot\mathbf{a}_i(\hat\phi) \neq 0$, the same argument as in the proof of (1) for the normal transonic state can be applied to prove that the right hand side of the preceding equation approaches positive infinity as $v \uparrow v_L$. □

Proof of necessity in Theorem 3.28 We recall that the trace of a square matrix is equal to the sum of its eigenvalues. In the case where the transonic state is normal, (3.140) follows from (1) of Lemma 3.31 and the increasing monotonicity of each eigenvalue of $\mathbf{S}_3(v)$ as a function of v.

In the case where the transonic state is exceptional, (1) and (2) of Lemma 3.31 imply that $\mathbf{S}_3(v)$ and $\mathbf{S}_1(v)$ remain bounded in the transonic limit $v \uparrow v_L$. Hence formula (3.50) remains valid when $v \uparrow v_L$. The vector $\mathbf{l}(\hat\phi)$ which pertains to the traction on the surface $\mathbf{n}\cdot\mathbf{x} = 0$ produced by the limiting wave solution whose polarization is $\mathbf{a}(\hat\phi)$, namely

$$\mathbf{l}(\hat\phi) = [\mathbf{R}(0)^T + \tan\hat\phi\,\mathbf{T}(0)]\mathbf{a}(\hat\phi),$$

vanishes in the case where the transonic state is exceptional (see (3.136)). Therefore, by letting $v \uparrow v_L$ in (3.50) we get

$$\mathbf{S}_3(v_L)\mathbf{a}(\hat\phi) = \mathbf{0}.$$

This implies that the null space of $\mathbf{S}_3(v_L)$ is not trivial, so at least one eigenvalue of $\mathbf{S}_3(v)$ vanishes at $v = v_L$.

Suppose that the Rayleigh wave exists when the transonic state is exceptional. Theorem 3.12 implies that at least two eigenvalues of $\mathbf{S}_3(v)$ become zero at the same $v = v_R$ ($v_R < v_L$). Then by the increasing monotonicity of each eigenvalue of $\mathbf{S}_3(v)$ as a function of v, the two eigenvalues of $\mathbf{S}_3(v_L)$ must be positive, while the remaining eigenvalue of $\mathbf{S}_3(v_L)$ vanishes as just asserted. Therefore, we obtain (3.140). □

From (1) of Lemma 3.31 and the proof of sufficiency in Theorem 3.28 we immediately have another criterion for the existence of Rayleigh waves.

 Springer

Proposition 3.32 *When the transonic state is normal, the Rayleigh wave exists.*[52]

From this proposition and Definition 3.30 we come to the conclusion that for given orthogonal unit vectors **m** and **n**, a Rayleigh wave or a limiting body wave which produces no tractions on the surface exists [7, 10].

Note Limiting body waves in an isotropic material appear when $p_1 = \tan \hat{\phi} = 0$ (see the example after Remark 3.26). It follows from (3.158) in Exercise 3-2 that the vectors pertaining to the traction produced by the limiting wave solutions are

$$\mathbf{l}_1 = \mathbf{0}, \qquad \mathbf{l}_2 = \mu \mathbf{m} \neq \mathbf{0}.$$

Hence in isotropic elasticity, the transonic state is normal for any **m** and **n**.

3.6.3 Existence Criterion Based on **Z**

The surface impedance matrix $\mathbf{Z} = \mathbf{Z}(v)$ $(0 \leq v < v_L)$ defined by (3.41), or equivalently by (3.42), has the following properties:

- $\mathbf{Z}(v)$ is Hermitian for $0 \leq v < v_L$ (Corollary 3.10).
- $\mathbf{Z}(0)$ is positive definite (see the comment after Theorem 1.21).
- The Hermitian matrix $\frac{d}{dv}\mathbf{Z}(v)$ is negative definite for $0 < v < v_L$.

The last statement was shown in [41] indirectly by using the relation between the Lagrangian and the kinetic energy. Direct proofs are given in recent works [36, 48].
   A parallel method to the proof of (2) of Lemma 3.15 can be applied to obtain

Lemma 3.33 *For* $0 \leq v < v_L$, *the eigenvalues of* $\mathbf{Z} = \mathbf{Z}(v)$ *are monotonic decreasing functions of* v.

Proof Let $\mathbf{w}_i = \mathbf{w}_i(v) \in \mathbb{C}^3$ $(i = 1, 2, 3)$ be eigenvectors of $\mathbf{Z}$ associated with the eigenvalues $v_i = v_i(v) \in \mathbb{R}$ $(i = 1, 2, 3)$. Differentiating the eigenrelations

$$\mathbf{Z}\mathbf{w}_i = v_i\mathbf{w}_i, \qquad i = 1, 2, 3 \tag{3.154}$$

with respect to v, we have

$$\frac{d\mathbf{Z}}{dv}\mathbf{w}_i + \mathbf{Z}\frac{d\mathbf{w}_i}{dv} = \frac{dv_i}{dv}\mathbf{w}_i + v_i\frac{d\mathbf{w}_i}{dv}, \qquad i = 1, 2, 3.$$

Taking the inner products with $\mathbf{w}_i$, we get

$$\left(\mathbf{w}_i, \frac{d\mathbf{Z}}{dv}\mathbf{w}_i\right)_{\mathbb{C}^3} + \left(\mathbf{w}_i, \mathbf{Z}\frac{d\mathbf{w}_i}{dv}\right)_{\mathbb{C}^3} = \frac{dv_i}{dv}(\mathbf{w}_i, \mathbf{w}_i)_{\mathbb{C}^3} + v_i\left(\mathbf{w}_i, \frac{d\mathbf{w}_i}{dv}\right)_{\mathbb{C}^3}, \qquad i = 1, 2, 3.$$

Since $\mathbf{Z}$ is Hermitian, by (3.154) we have

$$\left(\mathbf{w}_i, \mathbf{Z}\frac{d\mathbf{w}_i}{dv}\right)_{\mathbb{C}^3} = \left(\mathbf{Z}\mathbf{w}_i, \frac{d\mathbf{w}_i}{dv}\right)_{\mathbb{C}^3} = v_i\left(\mathbf{w}_i, \frac{d\mathbf{w}_i}{dv}\right)_{\mathbb{C}^3}, \qquad i = 1, 2, 3.$$

---

[52]This proposition can be proved even when the outer profile of the slowness section has zero curvature at the point of tangency with L (see Exercise 3-4).

Then it follows that

$$\left(\mathbf{w}_i, \frac{d\mathbf{Z}}{dv}\mathbf{w}_i\right)_{\mathbb{C}^3} = \frac{dv_i}{dv}(\mathbf{w}_i, \mathbf{w}_i)_{\mathbb{C}^3}, \qquad i = 1, 2, 3.$$

The left hand side is negative because of the property of $\mathbf{Z}$ given as the third bullet above. Also, we have $(\mathbf{w}_i, \mathbf{w}_i)_{\mathbb{C}^3} > 0$ $(i = 1, 2, 3)$. Thus,

$$\frac{dv_i}{dv} < 0, \qquad i = 1, 2, 3.$$

This proves the lemma.   □

Since the positive definiteness of $\mathbf{Z}(0)$ implies that all the eigenvalues of $\mathbf{Z}(0)$ are positive, it follows from this lemma that all the eigenvalues of $\mathbf{Z}(v)$ decrease monotonically with v in the interval $0 \leq v < v_L$ from their positive values at $v = 0$.

Thus, if there exists v in the subsonic range $0 < v < v_L$ such that the eigenvalue of $\mathbf{Z}(v)$ vanishes, then Theorem 3.13 implies that this v is the velocity of the Rayleigh waves v_R and, by Theorem 3.14, it is unique.

Before proceeding to the existence criterion of Rayleigh waves, we shall see that $\mathbf{Z}(v)$ is well behaved in the transonic limit $v \uparrow v_L$. Namely, unlike the divergent behavior of $\mathbf{S}_3(v)$ as $v \uparrow v_L$ in the case where the transonic state is normal, $\mathbf{Z}(v)$ is bounded when $v \uparrow v_L$. In addition, the proof is much simpler than that of Lemma 3.31.

Proposition 3.34 *All the eigenvalues of* $\mathbf{Z}(v)$ *remain bounded as* $v \uparrow v_L$.

Proof Suppose that an eigenvalue of $\mathbf{Z}(v)$ diverges when $v \uparrow v_L$. Lemma 3.33 implies that this eigenvalue tends to negative infinity as $v \uparrow v_L$. Moreover, because all the eigenvalues of $\mathbf{Z}(v)$ are monotonic decreasing functions of v, and because the trace of the matrix is equal to the sum of its eigenvalues, $\mathrm{tr}\,\mathbf{Z}(v)$ must approach negative infinity as $v \uparrow v_L$.

On the other hand, by (3.45), the trace of the matrix $\mathbf{S}_2^{-1}\mathbf{S}_1$ vanishes for $0 \leq v < v_L$. Hence, taking the traces of both sides of (3.42) gives

$$\mathrm{tr}\,\mathbf{Z} = \mathrm{tr}\,\mathbf{S}_2^{-1} \qquad (0 \leq v < v_L). \qquad (3.155)$$

This is positive, because by Lemma 3.5, $\mathbf{S}_2(v)$, and hence $\mathbf{S}_2(v)^{-1}$, are positive definite for $0 \leq v < v_L$. Therefore, a contradiction arises.   □

Now we give the criteria in terms of $\mathbf{Z}(v)$ for the existence of Rayleigh waves.

Theorem 3.35 *For given orthogonal unit vectors* $\mathbf{m}$ *and* $\mathbf{n}$ *in* $\mathbb{R}^3$, *let* $\mathbf{Z} = \mathbf{Z}(v)$ *be the surface impedance matrix in* (3.41), *or equivalently in* (3.42), *and let* $v_L = v_L(\mathbf{m}, \mathbf{n})$ *be the limiting velocity given by* (3.20) *(or equivalently by* (3.132)). *Then the following three conditions are equivalent.*

(1)   *There exists a unique Rayleigh wave in the half-space* $\mathbf{n} \cdot \mathbf{x} \leq 0$ *which propagates along the surface* $\mathbf{n} \cdot \mathbf{x} = 0$ *in the direction of* $\mathbf{m}$ *with the phase velocity in the subsonic range* $0 < v < v_L$.

(2) *Either*

  (i) det $\mathbf{Z}(v_L) < 0$, *or*
  (ii) $2(v_1 v_2 + v_2 v_3 + v_3 v_1)(v_L) = (\operatorname{tr} \mathbf{Z}(v_L))^2 - \operatorname{tr} \mathbf{Z}(v_L)^2 < 0$,

  where

$$\mathbf{Z}(v_L) = \lim_{v \uparrow v_L} \mathbf{Z}(v)$$

  and $v_i(v)$ $(i = 1, 2, 3)$ *are the eigenvalues of* $\mathbf{Z}(v)$.
(3) *There exists an* $\varepsilon > 0$ *such that*

$$\det \mathbf{Z}(v) < 0 \quad for \quad v_L - \varepsilon < v < v_L.$$

To prove this theorem, the following lemma on the behavior of the eigenvalues of $\mathbf{Z}(v_L)$ is helpful.

Lemma 3.36 (1) *At most one eigenvalue of* $\mathbf{Z}(v)$ *can be negative at* $v = v_L$.
(2) *If one of the eigenvalues of* $\mathbf{Z}(v)$ *is negative at* $v = v_L$, *then at least one of the others is positive at* $v = v_L$.

Proof (1) Let $\mathbf{w}_i = \mathbf{w}_i(v) \in \mathbb{C}^3$ $(i = 1, 2, 3)$ be a set of orthonormal eigenvectors of the Hermitian matrix $\mathbf{Z}(v)$ associated with the eigenvalues $v_i = v_i(v) \in \mathbb{R}$ $(i = 1, 2, 3)$. Suppose that two eigenvalues $v_1(v)$ and $v_2(v)$ are negative at $v = v_L$. We take a non-zero vector $\mathbf{b} \in \mathbb{R}^3$ such that $(\mathbf{b}, \mathbf{w}_3(v_L))_{\mathbb{C}^3} = 0$. Then for such $\mathbf{b}$, it follows from (3.65) and (3.144) that

$$(\mathbf{b}, \mathbf{Z}(v_L)\mathbf{b})_{\mathbb{C}^3} = \left(\mathbf{b}, \left(\sum_{i=1}^{3} v_i(v_L)\, \mathbf{w}_i(v_L) \otimes \mathbf{w}_i(v_L)\right)\mathbf{b}\right)_{\mathbb{C}^3}$$

$$= \sum_{i=1}^{3} v_i(v_L)\, (\mathbf{b}, \mathbf{w}_i(v_L))_{\mathbb{C}^3}\, \overline{(\mathbf{b}, \mathbf{w}_i(v_L))_{\mathbb{C}^3}}$$

$$= \sum_{i=1}^{2} v_i(v_L)\, (\mathbf{b}, \mathbf{w}_i(v_L))_{\mathbb{C}^3}\, \overline{(\mathbf{b}, \mathbf{w}_i(v_L))_{\mathbb{C}^3}},$$

which is negative from the supposition.

On the other hand, recalling that $\mathbf{S}_2^{-1}\mathbf{S}_1$ is anti-symmetric for $0 \leq v < v_L$, we obtain from (3.42)

$$(\mathbf{b}, \mathbf{Z}(v)\mathbf{b})_{\mathbb{C}^3} = \mathbf{b} \cdot \mathbf{S}_2^{-1}(v)\mathbf{b} \qquad (0 \leq v < v_L).$$

This is positive, because $\mathbf{S}_2^{-1}(v)$ is positive definite for $0 \leq v < v_L$. Therefore, by taking the limit $v \uparrow v_L$ it follows that $(\mathbf{b}, \mathbf{Z}(v_L)\mathbf{b})_{\mathbb{C}^3}$ is non-negative. This is a contradiction.
(2) Recalling that the right hand side of (3.155) is positive and that the trace of a square matrix is equal to the sum of its eigenvalues, we obtain the assertion by taking the limit $v \uparrow v_L$ on both sides of (3.155). □

Remark 3.37 Assertion (1) of Lemma 3.36, when combined with Lemma 3.33 and Theorem 3.13, immediately implies that the Rayleigh wave is unique if it exists. This is an alternative proof of Theorem 3.14 by the approach using $\mathbf{Z}(v)$.

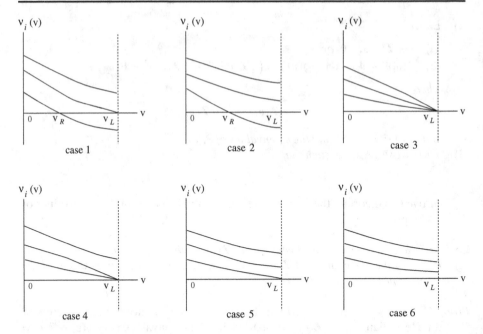

Fig. 2 Variation of eigenvalues of $\mathbf{Z}(v)$

Proof of Theorem 3.35 On the basis of the preceding lemma we can illustrate how the three eigenvalues $v_i = v_i(v)$ $(i = 1, 2, 3)$ of $\mathbf{Z}(v)$ decrease with v in the interval $0 \leq v \leq v_L$ from their positive values at $v = 0$. Then we classify the variations of these eigenvalues into the six cases depending on whether each eigenvalue is positive, zero, or negative at $v = v_L$ (see Fig. 2).

Rayleigh waves exist in the Cases 1 and 2. Looking at the figure carefully and noting that the determinant of a square matrix is equal to the product of all its eigenvalues, we obtain the result.[53] □

Example Let us compute $\mathbf{Z}(v)$ of an isotropic material for $0 \leq v \leq v_L$. From (3.36) it follows that

$$\det[\mathbf{a}_1, \mathbf{a}_2, \mathbf{a}_3] = -1 - p_1 p_3$$

and

$$\mathrm{Cof}\,[\mathbf{a}_1, \mathbf{a}_2, \mathbf{a}_3] = \left[-\ell(1 + p_1 p_3), \ p_3 \mathbf{m} - \mathbf{n}, \ -\mathbf{m} - p_1 \mathbf{n}\right], \qquad (3.156)$$

where $\ell = \mathbf{m} \times \mathbf{n}$ (Exercise 3-5). Then simple algebraic calculations of

$$\mathbf{Z}(v) = -\sqrt{-1}\,[\mathbf{l}_1, \mathbf{l}_2, \mathbf{l}_3][\mathbf{a}_1, \mathbf{a}_2, \mathbf{a}_3]^{-1} = \sqrt{-1}\,\frac{1}{1 + p_1 p_3}\,[\mathbf{l}_1, \mathbf{l}_2, \mathbf{l}_3]\,\mathrm{Cof}\,[\mathbf{a}_1, \mathbf{a}_2, \mathbf{a}_3]^{T},$$

---

[53] It is proved that at least one eigenvalue of $\mathbf{Z}(v_L)$ is positive and that all three eigenvalues of $\mathbf{Z}(v_L)$ cannot be positive ([12]). These prohibit the variation of $v_i(v)$ $(i = 1, 2, 3)$ depicted in the cases 3 and 6, but do not affect the results of the theorem.

combined with (3.158) in Exercise 3-2, lead to

$$\mathbf{Z}(v) = -\sqrt{-1}\Bigg[\mu\, p_1 \boldsymbol{\ell} \otimes \boldsymbol{\ell} + \frac{V\, p_3}{1 + p_1 p_3}\mathbf{m} \otimes \mathbf{m} + \frac{V\, p_1}{1 + p_1 p_3}\mathbf{n} \otimes \mathbf{n}$$

$$+ \left(2\mu - \frac{V}{1 + p_1 p_3}\right)(\mathbf{m} \otimes \mathbf{n} - \mathbf{n} \otimes \mathbf{m})\Bigg], \qquad (3.157)$$

where $V = \rho\, v^2$ (Exercise 3-5).

When $v = v_L^{\mathrm{Iso}} = \sqrt{\frac{\mu}{\rho}}$, it holds that

$$p_1 = 0, \qquad p_3 = \sqrt{-1}\sqrt{\frac{\lambda + \mu}{\lambda + 2\mu}}.$$

Hence

$$\mathbf{Z}\!\left(v_L^{\mathrm{Iso}}\right) = \mu\sqrt{\frac{\lambda + \mu}{\lambda + 2\mu}}\,\mathbf{m} \otimes \mathbf{m} - \sqrt{-1}\,\mu(\mathbf{m} \otimes \mathbf{n} - \mathbf{n} \otimes \mathbf{m}),$$

whose eigenvalues are 0 and the two real roots of the quadratic equation of p

$$p^2 - \mu\sqrt{\frac{\lambda + \mu}{\lambda + 2\mu}}\, p - \mu^2 = 0.$$

Obviously, this has two roots of opposite signs. Therefore, isotropic elasticity falls under Case 1.

We shall consider the existence of Rayleigh waves for transversely isotropic materials and for orthorhombic materials in Exercises 3-7 and 3-8.

3.6.4 Existence Criterion Based on Slowness Sections

Let us turn to the slowness section in the **m-n** plane (see Definition 3.24 and Proposition 3.25). The vertical line L approaching the slowness section from the right can first touch the outer profile tangentially at one, two or three points and two or all three of the constituent closed curves may be tangent to one another at a point of contact (cf. Remark 3.26). Chadwick and Smith [22] classified the transonic states into six types:

Type 1. L touches the outer profile once at one point belonging to one closed curve of the slowness section.

Type 2. L touches the outer profile once at one point belonging to two closed curves of the slowness section.

Type 3. L touches the outer profile once at one point belonging to all three closed curves of the slowness section.

Type 4. L touches the outer profile twice, each point of contact belonging to a single closed curve of the slowness section.

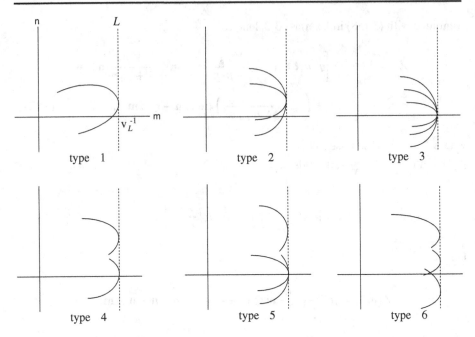

Fig. 3 Six types of transonic states

Type 5. L touches the outer profile twice, one point of contact belonging to a
single closed curve of the slowness section and the other point of contact
belonging to two closed curves.

Type 6. L touches the outer profile thrice, each point of contact belonging to just
one closed curve of the slowness section.

These possibilities are illustrated in Fig. 3.[54]

For instance, isotropic elasticity falls under Type 2 (see the example immediately
following Remark 3.26).

Now we give an existence criterion for Rayleigh waves based on the above
classification of the transonic state.

Theorem 3.38 *In the* **m-n** *plane, assume that the outer profile of the slowness section
has non-zero curvature at the point of tangency with L. Then the Rayleigh waves in
the half-space* $\mathbf{n} \cdot \mathbf{x} \leq 0$ *which propagate along the surface* $\mathbf{n} \cdot \mathbf{x} = 0$ *in the direction
of* **m** *with the phase velocity in the subsonic range* $0 < v < v_L$ *always exist when the
transonic state is of Type 2, 3, 4, 5, or 6.*

---

[54] A lucid illustration containing the whole closed curves in the slowness sections is given in Section
6.C of [22]. Each type of the transonic state corresponds to a different scenario regarding limiting
body waves. For example, in Type 5 three limiting body waves arise, two of which have the same
direction of propagation but have mutually orthogonal polarizations. For interpretation of all six
types above, we refer to [22].

The theorem is proved in [12, 22] by showing that transonic states of types 2, 3, 4, 5, and 6 are necessarily normal in the sense of Definition 3.30.

Remark 3.39 For a transonic state of Type 1, there is no longer a geometric criterion for the existence of Rayleigh waves in terms of the slowness section. Hence we must use the other criteria in the preceding subsections for this state.

The case where the outer profile of the slowness section has zero curvature at the point of tangency with L is examined in [13]. It is proved there that Rayleigh waves always exist in this case.

3.7 Comments and References

The arguments in Sections 3.1, 3.2, 3.3, and 3.4 follow essentially the theory of surface waves summarized in Chapters 6 to 8 of [22]. The existence of surface waves for which Stroh's eigenvalue problem (3.14) is degenerate is considered in [37, 79, 83, 86]. For supersonic surface waves, we refer, for example, to [32, 44] and the references therein. Theorem 3.12, which we have used to find the secular equations for Rayleigh waves in Sections 3.4 and 3.5, goes back to the work in [14]. The proof of Lemma 3.15 is given in [21]. The uniqueness of the Rayleigh wave is proved in [12, 14, 22]. A relevant review on the studies of wave propagation with the use of the Stroh formalism, which covers Rayleigh waves, interfacial waves, body waves and supersonic surface waves, is given by [7].

We have given two methods for finding the equation (3.82) for v_R^{Iso}; one method uses rank $\mathbf{S}_3\left(v_R^{\text{Iso}}\right) = 1$ (Section 3.4) and the other uses $\det[\mathbf{l}_1, \mathbf{l}_2, \mathbf{l}_3] = 0$ $\left(v = v_R^{\text{Iso}}\right)$ (Exercise 3-2). However, the most traditional method is to decompose the surface-wave solution into its longitudinal and transversal components and to find the condition on the velocity so that this superposition produces no traction on the surface and decays with depth below the surface (see, for example, Section 5.3 of [3] and Section 3.24 of [38]). Of course, this method is applicable only to isotropic elasticity.

The usefulness of the matrix $\mathbf{S}_3(v)$ can be well recognized in the derivation of the formula for the phase velocity of Rayleigh waves in weakly anisotropic elastic media. Section 3.5 is based on [70, 71], except that the initial stress is set equal to zero. It is somewhat surprising that only four components of the perturbative part (3.91) can affect the first-order perturbation of the velocity. The perturbation formula in Theorem 3.19 and that in [71] do not agree totally with the result in [27]. In [71] an argument is given to support our present results.

The slowness surfaces and slowness sections are important tools in the theory of elastic wave propagation (see, for example, [50]). The arguments in Subsections 3.6.1 and 3.6.2 are based mainly on Chapters 6 and 9 of [22]. Subsection 3.6.3 follows essentially [12]. The arguments in [12] are based on the analysis of the surface impedance matrix $\mathbf{Z}(v)$, which are independent of whether the curvature of the outer profile of the slowness section vanishes at the point of tangency with L. The approach there also proves Proposition 3.32 and Theorem 3.38, and it proves the existence of Rayleigh waves when the outer profile of the slowness section has zero curvature at the point of tangency with L (cf. [12, 13]).

In this chapter, we have wanted to make our discussions as self-contained as possible. In fact, until Subsection 3.6.1 we need not appeal to the positive definiteness of $\mathbf{Z}(0)$, $-\mathbf{S}_3(0)$ and $-\frac{d\mathbf{Z}(v)}{dv}$ $(0 < v < v_L)$, the direct proofs of which are long and complicated.

It should be noted that Chadwick and Ting [24] proved the equivalence between the negative definiteness of $\mathbf{S}_3(v)$ and the positive definiteness of $\mathbf{Z}(v)$ for $0 \le v < v_L$. In fact, they showed that for any $\mathbf{u} \in \mathbb{R}^3$ and $\mathbf{v} = \mathbf{v}^+ + \sqrt{-1}\,\mathbf{v}^- \in \mathbb{C}^3$ $(\mathbf{v}^+, \mathbf{v}^- \in \mathbb{R}^3)$,

$$\mathbf{u} \cdot \left(-\mathbf{S}_3(v)\,\mathbf{u}\right) = A^{-1}F\{(\mathbf{u} \cdot \mathbf{a}')^2 + (\mathbf{u} \cdot \mathbf{b}')^2\} + C^{-1}(\mathbf{u} \cdot \mathbf{c}')^2,$$

$$\left(\mathbf{v}, \mathbf{Z}(v)\,\mathbf{v}\right)_{\mathbb{C}^3} = A^{-1}\Big[(\mathbf{v}^+ \cdot \mathbf{a}' - s\,\mathbf{v}^- \cdot \mathbf{b}')^2 + (\mathbf{v}^+ \cdot \mathbf{b}' + s\,\mathbf{v}^- \cdot \mathbf{a}')^2$$
$$+ F\{(\mathbf{v}^- \cdot \mathbf{a}')^2 + (\mathbf{v}^- \cdot \mathbf{b}')^2\}\Big] + C^{-1}\left\{(\mathbf{v}^+ \cdot \mathbf{c}')^2 + (\mathbf{v}^- \cdot \mathbf{c}')^2\right\}.$$

Here $\mathbf{a} \pm \sqrt{-1}\,\mathbf{b}$ $(\mathbf{a}, \mathbf{b} \in \mathbb{R}^3)$ and $\mathbf{c}$ are the eigenvectors of $\mathbf{S}_1(v)$ associated with the eigenvalues $\pm\sqrt{-1}\,s$ $(s > 0)$ and 0, respectively,

$$\mathbf{a}' = \frac{\mathbf{b} \times \mathbf{c}}{g}, \qquad \mathbf{b}' = \frac{\mathbf{c} \times \mathbf{a}}{g}, \qquad \mathbf{c}' = \frac{\mathbf{a} \times \mathbf{b}}{g}, \qquad g = \det[\mathbf{a}, \mathbf{b}, \mathbf{c}],$$

and

$$A = \mathbf{a}' \cdot \left(\mathbf{S}_2(v)\,\mathbf{a}'\right), \qquad C = \mathbf{c}' \cdot \left(\mathbf{S}_2(v)\,\mathbf{c}'\right), \qquad F = 1 + \frac{1}{2}\mathrm{tr}\,\mathbf{S}_1(v)^2.$$

From the above we obtain another secular equation

$$F = 1 + \frac{1}{2}\mathrm{tr}\,\mathbf{S}_1(v)^2 = 0,$$

which is used in [23].

A detailed study on body waves, transonic states and surface waves in transversely isotropic materials is given in [18]. Further investigations of surface waves polarized in a plane of material symmetry of monoclinic, orthorhombic, and cubic media can be found in [8, 19, 20]. Secular equations for v_R in monoclinic and general anisotropic materials are proposed by Ting [81, 82] on the basis of the Stroh formalism.

Nakamura [52] gave, in terms of the surface impedance matrix, a condition for the existence of Rayleigh waves for inhomogeneous elastic media with curved boundaries. His study is based on microlocal analysis combined with the Stroh formalism.

3.8 Exercises

3-1 Show that in isotropic elasticity the solutions p_α $(\alpha = 1, 2, 3, \ \mathrm{Im}\ p_\alpha > 0)$ to (3.11) and the displacement parts $\mathbf{a}_\alpha$ $(\alpha = 1, 2, 3)$ of eigenvectors of $\mathbf{N}$ are given by (3.35) and (3.36), respectively.

 Springer

3-2 Using (3.12), show that in isotropic elasticity the traction parts $\mathbf{l}_\alpha$ ($\alpha = 1, 2, 3$) of the eigenvectors $\begin{bmatrix} \mathbf{a}_\alpha \\ \mathbf{l}_\alpha \end{bmatrix}$ ($\alpha = 1, 2, 3$) of $\mathbf{N}$ where $\mathbf{a}_\alpha$ ($\alpha = 1, 2, 3$) are given by (3.36) are written as

$$\mathbf{l}_1 = \mu\, p_1 \mathbf{m} \times \mathbf{n}, \qquad \mathbf{l}_2 = \mu\left((1 - p_1^2)\mathbf{m} + 2p_1\mathbf{n}\right),$$
$$\mathbf{l}_3 = 2\mu\, p_3 \mathbf{m} + \left(\lambda + p_3^2(\lambda + 2\mu)\right)\mathbf{n}. \tag{3.158}$$

Then derive the secular equation (3.83) from (3.47).

3-3 By Theorem 3.12, the vanishing of the $(3, 2)$ minor of (3.98) would give another approximate secular equation for Rayleigh waves in weakly anisotropic elastic media in Section 3.5, which is written as

$$s_{23}(v) = 0. \tag{3.159}$$

(a) Prove that the effect of the perturbative part $\mathbf{A}$ in (3.91) on the $(2, 3)$ component of the integrand (3.105) comes from $a_{22}, a_{23}, a_{33}, a_{44}, a_{24}$ and a_{34}.

(b) Prove that the coefficients of a_{22}, a_{23}, a_{33} and a_{44} in the $(2, 3)$ components of (3.105) are odd functions in ϕ. Therefore, the terms linear in these components included in $s_{23}(v)$ vanish.

(c) Use some software for symbolic calculus (e.g., MATHEMATICA or MAPLE) to show that the coefficients of a_{24} and a_{34} in $s_{23}(v)$ vanish when $v = v_R^{\text{Iso}}$.

Therefore, equation (3.159) is simply an identity, to the first order in the perturbative part $\mathbf{A}$.

3-4 When the outer profile of the slowness section has zero curvature at the point of tangency with L, it follows that for $n = 4$ or 6,

$$\frac{d^k}{d\phi^k}\left(\frac{c_i(\phi)}{\cos\phi}\right) = 0, \quad k = 1, 2, \cdots, n - 1, \qquad \frac{d^n}{d\phi^n}\left(\frac{c_i(\phi)}{\cos\phi}\right) > 0$$
$$\text{at} \quad \phi = \hat{\phi}$$

(cf. [13]). Modify the first half of the proof of (1) of Lemma 3.31 to show that Proposition 3.32 holds in this case.

3-5 Check (3.156) and complete the calculations of (3.157).

3-6 Take the limit $v \downarrow 0$ in (3.157). Then show that the result coincides with (1.95).

3-7 For a transversely isotropic material whose axis of symmetry coincides with the 3-axis, consider surface waves which propagate along the surface of the half-space $x_2 \leq 0$ in the direction of the 1-axis.

(a) Under the notations (1.100), show that the limiting velocity v_L^{Trans} is given by

$$\rho\left(v_L^{\text{Trans}}\right)^2 = \min\left(L, \frac{A - N}{2}\right).$$

(b) Show that the surface impedance matrix is given by

$$\mathbf{Z}(v) = \left(Z_{ij}\right)_{i\downarrow j \to 1,2,3} = \overline{\mathbf{Z}(v)}^T,$$

$$Z_{11} = \frac{\sqrt{\frac{A-N}{2}(A-V)}}{\sqrt{A(A-V)} + \sqrt{\frac{A-N}{2}\left(\frac{A-N}{2} - V\right)}}$$

$$\times \left\{ \sqrt{A\left(\frac{A-N}{2} - V\right)} + \sqrt{\frac{A-N}{2}(A-V)} \right\},$$

$$Z_{22} = \frac{\sqrt{A\left(\frac{A-N}{2} - V\right)}}{\sqrt{A(A-V)} + \sqrt{\frac{A-N}{2}\left(\frac{A-N}{2} - V\right)}}$$

$$\times \left\{ \sqrt{A\left(\frac{A-N}{2} - V\right)} + \sqrt{\frac{A-N}{2}(A-V)} \right\},$$

$$Z_{12} = \frac{-\sqrt{-1}\sqrt{\frac{A-N}{2}}}{\sqrt{A(A-V)} + \sqrt{\frac{A-N}{2}\left(\frac{A-N}{2} - V\right)}}$$

$$\times \left\{ \sqrt{\frac{A-N}{2}A(A-V)} - N\sqrt{\frac{A-N}{2} - V} \right\},$$

$$Z_{13} = Z_{23} = 0, \qquad Z_{33} = \sqrt{L(L-V)},$$

where $V = \rho v^2$.

(c) Prove that Rayleigh waves which propagate along the surface of the half-space $x_2 \leq 0$ in the direction of the 1-axis exist if and only if L satisfies

$$\rho \left(v_R^{\text{Trans}}\right)^2 < L,$$

where $v = v_R^{\text{Trans}}$ is the solution to (3.83) with $V = \rho v^2$, $\lambda = N$ and $\mu = \frac{A-N}{2}$ in the range $0 < V < \frac{A-N}{2}$. Show that the phase velocity of the Rayleigh waves in question is equal to v_R^{Trans} if they exist.

(Cf. [12, 41] and Sections 8-A and 9-E of [22]. Supersonic surface waves which propagate along the surface of the half-space $x_2 \leq 0$ in the direction of the 1-axis exist when $L < \rho \left(v_R^{\text{Trans}}\right)^2$; cf. the references cited.)

3-8 For an orthorhombic material whose elasticity tensor $\mathbf{C}$ is given by (1.15), consider surface waves which propagate along the surface of the half-space $x_2 \leq 0$ in the direction of the 1-axis.

(a) Show that the limiting velocity v_L^{Orth} is given by

$$\rho \left(v_L^{\text{Orth}}\right)^2 = \min\left(C_{55}, V^*\right),$$

where

$$V^* = \sup_{0 < V \leq \min(C_{11}, C_{66})} \left\{ \sqrt{C_{22}(C_{11} - V)} + \sqrt{C_{66}(C_{66} - V)} \geq |C_{12} + C_{66}| \right\}.$$

(b) Show that the surface impedance matrix is given by

$$\mathbf{Z}(v) = \left(Z_{ij}\right)_{i\downarrow j \to 1,2,3} = \overline{\mathbf{Z}(v)}^T,$$

$$Z_{11} = \frac{\sqrt{C_{66}(C_{11} - V)}}{\sqrt{C_{22}(C_{11} - V)} + \sqrt{C_{66}(C_{66} - V)}}$$

$$\times \sqrt{\left(\sqrt{C_{22}(C_{11} - V)} + \sqrt{C_{66}(C_{66} - V)}\right)^2 - (C_{12} + C_{66})^2},$$

$$Z_{22} = \frac{\sqrt{C_{22}(C_{66} - V)}}{\sqrt{C_{22}(C_{11} - V)} + \sqrt{C_{66}(C_{66} - V)}}$$

$$\times \sqrt{\left(\sqrt{C_{22}(C_{11} - V)} + \sqrt{C_{66}(C_{66} - V)}\right)^2 - (C_{12} + C_{66})^2},$$

$$Z_{12} = \frac{-\sqrt{-1}\sqrt{C_{66}}}{\sqrt{C_{22}(C_{11} - V)} + \sqrt{C_{66}(C_{66} - V)}}$$

$$\times \left(\sqrt{C_{22}C_{66}(C_{11} - V)} - C_{12}\sqrt{C_{66} - V}\right),$$

$$Z_{13} = Z_{23} = 0, \qquad Z_{33} = \sqrt{C_{44}(C_{55} - V)},$$

where $V = \rho v^2$. (Cf. Section 12-10 of [77].)

(c) Show that Rayleigh waves which propagate along the surface of the half-space $x_2 \leq 0$ in the direction of the 1-axis exist if $V^* \leq C_{55}$. (Suggestion: The eigenvalues of $\mathbf{Z}(v)$ are Z_{33} and the two real roots of the following quadratic equation of p:

$$p^2 - (Z_{11} + Z_{22}) p + Z_{11} Z_{22} + Z_{12}^2 = 0.$$

The preceding equation has two roots of opposite signs if and only if $Z_{11} Z_{22} + Z_{12}^2 < 0$. Use Fig. 2 (Cf. [17]).)

(d) Show that the secular equation for the phase velocity of the Rayleigh waves in question is given by (3.122)[55] if they exist.

(e) Consider surface waves which propagate along the surface perpendicular to the axis of symmetry of a transversely isotropic material. Derive the formulas for the limiting velocity and the surface impedance matrix and find sufficient conditions for the existence of Rayleigh waves which propagate along the surface perpendicular to the axis of symmetry.

References

1. Akamatsu, M., Nakamura, G., Steinberg, S.: Identification of Lamé coefficients from boundary observations. Inverse Probl. **7**, 335–354 (1991)
2. Akamatsu, M., Tanuma, K.: Green's function of anisotropic piezoelectricity. Proc. R. Soc. Lond. A **453**, 473–487 (1997)

---

[55] Of course, since we consider here the Rayleigh waves which propagate along the surface of the half-space $x_2 \leq 0$ in the direction of the 1-axis, we must change the indices $1, 2, 3$ of C_{ijkl} in (3.123) to the indices $3, 1, 2$, respectively.

🕮 Springer

3. Aki, K., Richards, P.G.: Quantitative Seismology, 2nd edn. University Science Books (2002)
4. Alshits, V.I., Darinskii, A.N., Lothe, J.: On the existence of surface waves in half-infinite anisotropic elastic media with piezoelectric and piezomagnetic properties. Wave Motion **16**, 265–283 (1992)
5. Alshits, V.I., Kirchner, H.O.K., Ting, T.C.T.: Angularly inhomogeneous piezoelectric piezomagnetic magnetoelectric anisotropic media. Phil. Mag. Lett. **71**, 285–288 (1995)
6. Barnett, D.M.: The precise evaluation of derivatives of the anisotropic elastic Green's functions. Phys. Status Solidi, B **49**, 741–748 (1972)
7. Barnett, D.M.: Bulk, surface, and interfacial waves in anisotropic linear elastic solids. Int. J. Solids Struct. **37**, 45–54 (2000)
8. Barnett, D.M., Chadwick, P., Lothe, J.: The behaviour of elastic surface waves polarized in a plane of material symmetry. I. Addendum. Proc. R. Soc. Lond. A **433**, 699–710 (1991)
9. Barnett, D.M., Lothe, J.: Synthesis of the sextic and the integral formalism for dislocations, Green's functions, and surface waves in anisotropic elastic solids. Phys. Norv. **7**, 13–19 (1973)
10. Barnett, D.M., Lothe, J.: Consideration of the existence of surface wave (Rayleigh wave) solutions in anisotropic elastic crystals. J. Phys. F: Metal Phys. **4**, 671–686 (1974)
11. Barnett, D.M., Lothe, J.: Dislocations and line charges in anisotropic piezoelectric insulators. Phys. Status Solidi, B **67**, 105–111 (1975)
12. Barnett, D.M., Lothe, J.: Free surface (Rayleigh) waves in anisotropic elastic half spaces: the surface impedance method. Proc. R. Soc. Lond. A **402**, 135–152 (1985)
13. Barnett, D.M., Lothe, J., Gunderson, S.A.: Zero curvature transonic states and free surface waves in anisotropic elastic media. Wave Motion **12**, 341–360 (1990)
14. Barnett, D.M., Lothe, J., Nishioka, K., Asaro, R.J.: Elastic surface waves in anisotropic crystals: a simplified method for calculating Rayleigh velocities using dislocation theory. J. Phys. F: Metal Phys. **3**, 1083–1096 (1973)
15. Brown, R.M.: Recovering the conductivity at the boundary from the Dirichlet to Neumann map: a pointwise result. J. Inverse Ill-posed Probl. **9**, 567–574 (2001)
16. Calderón, A.P.: On an inverse boundary value problem. In: Seminar on Numerical Analysis and its Applications to Continuum Physics, pp. 65–73. Soc. Brasileira de Matemática, Rio de Janeiro (1980)
17. Chadwick, P.: The existence of pure surface modes in elastic materials with orthorhombic symmetry. J. Sound Vib. **47**, 39–52 (1976)
18. Chadwick, P.: Wave propagation in transversely isotropic elastic media I. Homogeneous plane waves. II. Surface waves. III. The special case $a_5 = 0$ and the inextensible limit. Proc. R. Soc. Lond. A **422**, 23–66, 67–101, 103–121 (1989)
19. Chadwick, P.: The behaviour of elastic surface waves polarized in a plane of material symmetry I. General analysis. Proc. R. Soc. Lond. A **430**, 213–240 (1990)
20. Chadwick, P., Wilson, N.J.: The behaviour of elastic surface waves polarized in a plane of material symmetry II. Monoclinic media. III. Orthorhombic and cubic media. Proc. R. Soc. Lond. A **438**, 207–223, 225–247 (1992)
21. Chadwick, P., Jarvis, D.A.: Surface waves in a pre-stresses elastic body. Proc. R. Soc. Lond. A **366**, 517–536 (1979)
22. Chadwick, P., Smith, G.D.: Foundations of the theory of surface waves in anisotropic elastic materials. Adv. Appl. Mech. **17**, 303–376 (1977)
23. Chadwick, P., Smith, G.D.: Surface waves in cubic elastic materials, In: Hopkins, H.G., Sewell, M.J. (eds.) Mechanics of Solids, The Rodney Hill 60th Anniversary Volume, pp. 47–100. Pergamon, Oxford (1982)
24. Chadwick, P., Ting, T.C.T.: On the structure and invariance of the Barnett-Lothe tensors. Q. Appl. Math. **45** 419–427 (1987)
25. Chung, M.Y., Ting, T.C.T.: The Green function for a piezoelectric piezomagnetic magnetoelectric anisotropic medium with an elliptic hole or rigid inclusion. Phila. Mag. Lett. **72**, 405–410 (1995)
26. Ciarlet, P.G.: Mathematical Elasticity, Vol.1: Three-dimensional Elasticity. North-Holland, Amsterdam (1988)
27. Delsanto, P.P., Clark Jr., A.V.: Rayleigh wave propagation in deformed orthotropic materials. J. Acoust. Soc. Am. **81**, 952–960 (1987)
28. Duvaut, G., Lions, J.L.: Inequalities in Mechanics and Physics. Springer-Verlag, Berlin (1976)
29. Gantmacher, F.R.: The Theory of Matrices. American Mathematical Society, Providence, Rhode Island (1960)
30. Grisvard, P.: Elliptic Problems in Nonsmooth Domains. Pitman, London (1985)

31. Gunderson, S.A., Barnett, D.M., Lothe, J.: Rayleigh wave existence theory: a supplementary remark. Wave Motion **9**, 319–321 (1987)
32. Gunderson, S.A., Wang, L., Lothe, J.: Secluded supersonic elastic surface waves. Wave Motion **14**, 129–143 (1991)
33. Gurtin, M.E.: The linear theory of elasticity. In: Truesdell, C. (ed.) Handbuch der Physik, Vol. VIa/2, Mechanica of Solids II, pp. 1–295. Springer, Berlin (1972)
34. Hoger, A.: On the determination of residual stress in an elastic body. J. Elast. **16**, 303–324 (1986)
35. Huang, M., Man, C.-S.: Constitutive relation of elastic polycrystal with quadratic texture dependence. J. Elast. **72**, 183–212 (2003)
36. Ito, H.: Construction and existence criterion of Rayleigh and Stoneley waves by means of factorization of matrix polynomials (preprint)
37. Kosevich, A.M., Kosevich, Yu.A., Syrkin, E.S.: Generalized Rayleigh waves and the geometry of isofrequency surfaces of sound oscillation waves in crystals. Sov. Phys. JETP **61**, 639–644 (1985)
38. Landau, L.D., Lifshitz, E.M.: Theory of Elasticity, 2nd edn. Pergamon Press Ltd. (1970)
39. Lifshitz, I.M., Rozentsweig, L.N.: Construction of the Green tensor for the basic equation of the theory of elaticity for the case of an anisotropic infinite medium. Z. Eksp. Teor. Fiz. **17**, 783–791 (1947)
40. Lothe, J.: Dislocations in anisotropic media. In: Indenbom, V.L., Lothe, J. (eds.) Elastic Strain Fields and Dislocation Mobility, pp. 269-328. North-Holland, Amsterdam (1992)
41. Lothe, J., Barnett, D.M.: On the existence of surface-wave solutions for anisotropic elastic half-spaces with free surface. J. Appl. Phys. **47**, 428–433 (1976)
42. Lothe, J., Barnett, D.M.: Integral formalism for surface waves in piezoelectric crystals. Existence considerations. J. Appl. Phys. **47**, 1799–1807 (1976)
43. Lothe, J., Barnett, D.M.: Further development of the theory for surface waves in piezoelectric crystals. Phys. Norv. **8**, 239–254 (1976)
44. Lothe, J., Wang, L.: Self-orthogonal sextic formalism for anisotropic elastic media: spaces of simple reflection and two component surface waves. Wave Motion **21**, 163–181 (1995)
45. Malén, K.: A unified six-dimensional treatment of elastic Green's functions and dislocations. Phys. Status Solidi, B **44**, 661–672 (1971)
46. Man, C.-S., Carlson, D.E.: On the traction problem of dead loading in linear elasticity with initial stress. Arch. Ration. Mech. Anal. **128**, 223–247 (1994)
47. Man, C.-S., Lu, W.Y.: Towards an acoustoelastic theory for measurement of residual stress. J. Elast. **17**, 159–182 (1987)
48. Mielke, A., Fu, Y.B.: Uniqueness of the surface-wave speed: a proof that is independent of the Stroh formalism. Math. Mech. Solids **9**, 5–15 (2004)
49. Mura, T.: Micromechanics of Defects in Solids. Martinus Nijhoff, Dordrecht (1987)
50. Musgrave, M.J.P.: Crystal Acoustics. Holden-Day, San Fransisco (1970)
51. Nachman, A.I.: Global uniqueness for a two-dimensional inverse boundary value problem. Ann. Math. **142**, 71–96 (1995)
52. Nakamura, G.: Existence and propagation of Rayleigh waves and pulses. In: Wu, J.J., Ting, T.C.T., Barnett, D.M. (eds.) Modern Theory of Anisotropic Elasticity and Applications. SIAM Proceedings, pp. 215–231. SIAM, Philadelphia (1991)
53. Nakamura, G., Tanuma, K.: A nonuniqueness theorem for an inverse boundary value problem in elasticity. SIAM J. Appl. Math. **56**, 602–610 (1996)
54. Nakamura, G., Tanuma, K.: A formula for the fundamental solution of anisotropic elasticity. Q. J. Mech. Appl. Math. **50**, 179–194 (1997)
55. Nakamura, G., Tanuma, K.: Local determination of conductivity at the boundary from the Dirichlet-to-Neumann map. Inverse Problems **17**, 405–419 (2001)
56. Nakamura, G., Tanuma, K.: Reconstruction of the elasticity tensor of anisotropic elasticity at the boundary from the Dirichlet-to-Neumann map (in preparation).
57. Nakamura, G., Tanuma, K., Uhlmann, G.: Layer stripping for a transversely isotropic elastic medium. SIAM J. Appl. Math. **59**, 1879–1891 (1999)
58. Nakamura, G., Uhlmann, G.: Identification of Lamé parameters by boundary measurements. Am. J. Math. **115**, 1161–1187 (1993)
59. Nakamura, G., Uhlmann, G.: Global uniqueness for an inverse boundary value problem arising in elasticity. Invent. Math **118**, 457–474 (1994). Erratum, Invent. Math **152**, 205–207 (2003)
60. Nakamura, G., Uhlmann, G.: Inverse problems at the boundary for an elastic medium. SIAM J. Math. Anal. **26**, 263–279 (1995)
61. Nakamura, G., Uhlmann, G.: A layer stripping algorithm in elastic impedance tomography. In: Chavent, G., Papanicolaou, G., Sacks, P., Symes, W. (eds.) Inverse Problems in Wave

Propagation. The IMA Volumes in Mathematics and its Applications vol. 90, pp. 375–384. Springer, New York (1997)

62. Ristic, V.M.: Principles of Acoustic Devices. John Wiley & Sons, New York (1983)
63. Robertson, R.L.: Boundary identifiability of residual stress via the Dirichlet to Neumann map. Inverse Problems **13**, 1107–1119 (1997)
64. Royer, D., Dieulesaint, E.: Rayleigh wave velocity and displacement in orthorhombic, tetragonal, hexagonal, and cubic crystals. J. Acoust. Soc. Am. **76**, 1438–1444 (1984)
65. Stroh, A.N.: Dislocations and cracks in anisotropic elasticity. Phila. Mag. **3**, 625–646 (1958)
66. Stroh, A.N.: Steady state problems in anisotropic elasticity. J. Math. Phys. **41**, 77–103 (1962)
67. Sylvester, J., Uhlmann, G.: Inverse boundary value problems at the boundary-continuous dependence. Commun. Pure Appl. Math. **41**, 197–221 (1988)
68. Synge, J.L.: The Hypercircle in Mathematical Physics. Cambridge University Press, London (1957)
69. Tanuma, K.: Surface impedance tensors of transversely isotropic elastic materials. Q. J. Mech. Appl. Math. **49**, 29–48 (1996)
70. Tanuma, K., Man, C.-S.: Angular dependence of Rayleigh-wave velocity in prestressed polycrystalline media with monoclinic texture. J. Elast. **69**, 181–214 (2002)
71. Tanuma, K., Man, C.-S.: Perturbation formula for phase velocity of Rayleigh waves in prestressed anisotropic media. J. Elast. **85**, 21–37 (2006)
72. Tanuma, K., Man, C.-S., Huang, M., Nakamura, G.: Surface impedance tensors of textured polycrystals. J. Elast. **67**, 131–147 (2002)
73. Teutonico, L.J.: Dislocations in hexagonal crystals. Mater. Sci. Eng. **6**, 27–47 (1970)
74. Thurston, R.N.: Effective elastic coefficients for wave propagation in crystals under stress. J. Acoust. Soc. Am. **37**, 348–356 (1965). Erratum, J. Acoust. Soc. Am. **37**, 1147 (1965)
75. Tiersten, H.F.: Linear Piezoelectric Plate Vibrations. Plenum Press, New York (1969)
76. Ting, T.C.T.: Some identities and the structure of $\mathbf{N}_i$ in the Stroh formalism of anisotropic elasticity. Q. Appl. Math. **46**, 109–120 (1988)
77. Ting, T.C.T.: Anisotropic Elasticity. Oxford University Press, New York (1996)
78. Ting, T.C.T.: Existence of an extraordinary degenerate matrix **N** for anisotropic elastic materials. Q. J. Mech. Appl. Math. **49**, 405–417 (1996)
79. Ting, T.C.T.: Surface waves in anisotropic elastic materials for which the matrix $N(v)$ is extraordinary degenerate, degenerate, or semisimple. Proc. R. Soc. Lond. A **453**, 449–472 (1997)
80. Ting, T.C.T.: Recent developments in anisotropic elasticity. Int. J. Solids Struct. **37**, 401–409 (2000)
81. Ting, T.C.T.: Explicit secular equations for surface waves in monoclinic materials with the symmetry plane at $x_1 = 0$, $x_2 = 0$ or $x_3 = 0$. Proc. R. Soc. Lond. A **458**, 1017–1031 (2002)
82. Ting, T.C.T.: An explicit secular equation for surface waves in an elastic material of general anisotropy. Q. J. Mech. Appl. Math. **55**, 297–311 (2002)
83. Ting, T.C.T., Barnett, D.M.: Classification of surface waves in anisotropic elastic materials. Wave Motion **26**, 207–318 (1997)
84. Ting, T.C.T., Lee, V.-G.: The three-dimensional elastostatic Green's functions for general anisotropic linear elastic solids. Q. J. Mech. Appl. Math. **50**, 407–426 (1997)
85. Uhlmann, G.: Developments in inverse problems since Calderón's foundational paper. In: Harmonic Analysis and Partial Differential Equations. Chicago Lectures in Math., pp. 295–345. Univ. Chicago Press, Chicago IL (1999)
86. Wang, L.: Space of degeneracy in the Stroh eigensystem and surface waves in transversely isotropic elastic media. Wave Motion **40**, 173–190 (2004)

Index